Preparing for the Next Cyber Revolution

W0235261

Joseph N. Pelton

Preparing for the Next Cyber Revolution

How Our World Will Be Radically
Transformed – Again!

 Springer

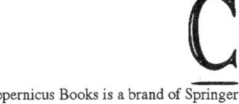

Copernicus Books is a brand of Springer

Joseph N. Pelton
International Association for the Advancement of Space Safety
Arlington, VA, USA

ISBN 978-3-030-02136-8   ISBN 978-3-030-02137-5   (eBook)
https://doi.org/10.1007/978-3-030-02137-5

Library of Congress Control Number: 2018961701

This Springer imprint is published by the registered company Springer Nature Switzerland AG
The registered company address is: Gewerbestrasse 11, 6330 Cham, Switzerland

Acknowledgments

This is a book that took a number of years to develop and evolve into its current form. Many people helped me to think about how the future is unfolding in today's world in such a chaotic and sometimes economic and politically unpredictable way. This book is not about small themes and issues. Indeed it is about fundamental issues such as: Can "democratic" values and systems survive the Internet? Can today's forms of capitalism and employment survive "super automation"? These issues demand a great thought and consideration by politicians, economists, technologists, and indeed everyone in society that values a stable, peaceful, and prosperous world that can escape the scourges of war and suppression of democratic values.

Colleagues and friends such as Professor Ram Jakhu of Canada, Dr. Scott Madry of North Carolina, and Dr. Peter Martinez of South Africa, among others have helped to clarify my thinking on the difficult and challenging topics addressed in this book. My friend, frequent coauthor, and superb editor Peter Marshall helped with not only the research, final editing, and structure of this book but also helped to dissect the critical thinking. Others have helped to inspire key thoughts in this book through their seminal writings. In this respect, I would particularly like to thank environmentalist Timothy Morton whose concept of "hyper objects" is a significant idea. It helps to explain why coping with climate change, technological innovation, and super automation is such a difficult problem to cope with in modern society. With sincerest thanks to all.

Arlington, VA, USA   Joseph N. Pelton

Contents

The original version of this book was revised. The correction is available at
https://doi.org/10.1007/978-3-030-02137-5_11

About the Author

Joseph N. Pelton Ph.D., is the former Dean and Chairman of the Board of Trustees of the International Space University. He also is the founder of the Arthur C. Clarke Foundation and the founding president of the Society of Satellite Professionals International – now known as the Space and Satellite Professionals International (SSPI). Dr. Pelton currently serves on the Executive Board of the International Association for the Advancement of Space Safety. He is the Director Emeritus of the Space and Advanced Communications Research Institute (SACRI) at George Washington University where he also served as director of the accelerated Master's program in telecommunications and computers from 1998 to 2004. Previously he headed the Interdisciplinary Telecommunications Program at the University of Colorado Boulder. Dr. Pelton has also served as president of the International Space Safety Foundation and president of the Global Legal Information Network (GLIN). Earlier in his career, he held a number of executive and management positions at COMSAT and INTELSAT, the global satellite organization where he was Director of Strategic Policy.

Dr. Pelton has been a speaker on national media in the United States (PBS NewsHour, Public Radio's All Things Considered, ABC, and CBS) and internationally on BBC, CBC, and France 3. He has spoken before Congress, the United Nations, and delivered talks in over 40 countries around the world. His honors include the Sir Arthur Clarke, International Achievement Award of the British Interplanetary Society, the Arthur

C. Clarke Foundation Award, the ICA Educator's Award, the ISCe Excellence in Education Award, and being elected to the International Academy of Astronautics. Most recently, in 2017, he won the Da Vinci Award of the International Association for the Advancement of Space Safety and the Guardian Award of the Lifeboat Foundation.

Dr. Pelton is a member of the SSPI Hall of Fame, Fellow of the IAASS, and Associate Fellow of the AIAA. Pelton is a widely published author with some 50 books written, co-authored or co-edited. His *Global Talk* won the Eugene Emme Literature Award of the International Astronautics Association and was nominated for a Pulitzer Prize. His most recent books include *The New Gold Rush: The Riches of Space Beckon!*, *Global Space Governance: An International Study*, and the second edition of *The Handbook of Satellite Applications*.

As director of Project SHARE, while Director of Strategic Policy for INTELSAT, he played a key role in the launching of the Chinese National TV University that now is the world's largest tele-education program. He received his degrees from the University of Tulsa, New York University, and Georgetown University, where he received his doctorate.

Chapter 1
What Is the Cyber Revolution?

Within a few decades, machine intelligence will surpass human intelligence, leading to The Singularity – technological change so rapid and profound it represents a rupture in the fabric of human history.

Ray Kurzweil, The Law of Accelerating Returns (http://www. kurzweilai.net)

I really think there are two fundamental paths [for humans]: One path is we stay on Earth forever, and some eventual extinction event wipes us out…. The alternative is, become a spacefaring and multi-planetary species.

Elon Musk (Nick Stockton, "Elon Musk Announces His Plan to Colonize Mars and Save Humanity" Science, Sept. 27, 2016 https://www.wired.com/2016/09/elon-musk-colonize-mars/)

An Epochal Shift in Human History?

The world as we know it is changing. This is no small change but fundamental change. These are sea-changes that will alter the course of human history. There are now truly massive drivers of change afoot. These basic shifts are much larger than most people comprehend. Certainly, we are not prepared for the "Cyber Revolution" or transformations that 'Space 2.0' can bring about. These capabilities can open the door to what we call "the Fourth Wave Economy." In this Fourth Wave Economy work will be redefined. The Internet will challenge democracy. NewSpace systems can not only create a new space economy but also totally new capabilities that might allow us to create new space structures that could allow us to better cope with climate change or even terraform Mars so that its newly created atmosphere can sustain vegetation and life.

The original version of this chapter was revised. The correction to this chapter is available at https://doi.org/10.1007/978-3-030-02137-5_11

© Springer Nature Switzerland AG 2019
J. N. Pelton, *Preparing for the Next Cyber Revolution*,
https://doi.org/10.1007/978-3-030-02137-5_1

Earning a living will take on new meaning in this Fourth Wave, an economy that will be circular rather than disposable. New technologies and new social and political practices can at least enable us to have a chance to meet the challenges of our times. Oddly enough the drive to cope with the effects of climate change will not only be our biggest problem but will also represent an opportunity. Coping with atmospheric change on Earth and perhaps on Mars – could also be one of the largest drivers of economic growth. Change that once took millennia to occur will now come to our planet in decades.

Ray Kurzweil has heralded the coming breakthroughs in artificial intelligence a true mega-driver of change. He has dubbed the movement to create human-like AI the "Singularity." Alvin Toffler would likely have called it the force behind the "Next Wave." In the author's book *Megacrunch*, I suggested that change would be driven by several forces that include "super automation," population growth, climate change, and other accelerators of economic, social and political change. These accelerators of change will be almost oxymorons. On one hand these breakthroughs, like super automation, will create huge problems of human employment, compensation, and the need to put the brakes on human population growth. Yet these same breakthroughs may prove vital to coping with climate change, coping with the so-called 'Sixth Extinction' and other challenges of the 'Fourth Wave' Economy. Thus technological breakthroughs will be both the 'yin' and the 'yang' of our future during the rest of the twenty-first century.[1]

Regardless of what we call this next epochal change in human history we can look forward to fundamental and even radical change in every aspect of our lives. This is not something looming in the distant future, but something that will unfold with stunning rapidity within the next few decades. The new Fourth Wave economy will probably take full effect within the next half century. Disruptions to democratic values and processes, climate change challenges and disruptions related to super automation, and over population are indeed occurring now. It will not be as sudden as the transformation of humanity as presented by Arthur C. Clarke in his novel *Childhood's End*, when all youth in the world in a matter of days merge into a giant mind meld where they begam functioning together as a giant global brain. Yet this is the going to be the biggest shift in what people do with their collective lives that ever before in human history. Changes will include how we earn a living, how rapidly we procreate, and where and how we live. If we are lucky we will even stave off the Athropocene disruption of our biosphere that is spaceship Earth. The question is will humans be smart enough to face the demands of future and do so within a very short period of time? In terms this vital condensed period of the Cyber-Revolution time we have to pull off a lot major changes at an accelerated pace.

It has been 4 million years since the arrival of the Southern Ape Man on Planet Earth. If we look at this period of 4 million years as what might be called a Cosmic

---

[1] In truth, it turns out that 'technological fixes' to human social issues are quite hard to accomplish and when implemented lack staying power. Automobiles were touted to the London City Council around 1900 as a fix to massive heaps of horse manure and city pollution. More recently, the noted economist John Kenneth Galbraith in the 1950s wrote a book that was much celebrated at the time. It predicted a bright future fueled by technology. This book *The Affluent Society* that held out so much hope for the future is today considered an exercise in looking to the future through rose-colored glasses. (John Kenneth Galbraith, *The Affluent Society*, (1958) https://www.goodreads.com/book/show/41589.The_Affluent_Society)

Super Month, then every second of this cosmic time represent 2 years. On this cosmic scale the invention of agriculture comes 1.5 h before midnight of the 30th day of the month. The Renaissance comes 4 min to midnight, and the Industrial Revolution is just 2 min shy of midnight. We have less than a minute of Super Month Time to avoid the sixth extinction, control human population growth, cope with climate change and transform our political, economic and social systems in a way that represents a pathway to survival.

In short, we seem ill-prepared for all the changes that will affect all our lives. This sea-change in how and why we do almost everything will truly be revolutionary and change the entire race of *Homo sapiens* in many diverse ways. We will become a true space-faring species armed with powerful new technologies undreamed of only a few years ago. Smart robots will not only take over labors in agriculture, mining, and manufacturing but also in a rising spectrum of service occupations – in areas that we today consider to be the domain of humans, such as accounting, health services, pharmacology, engineering and more.

Why are we so very unprepared for this massive change to our lives? The answer apparently comes from environmental philosopher Timothy Morton, who invented the term "hyper-object" in 2007. Morton explained this concept of a hyper-object and what it meant in a book by that name in 2013. It is a term he used to describe a concept so large and so "massively distributed in time and space relative to humans" that we really have a hard time grasping it or its implications for our past, present, and especially our future.[2]

The universe, second-generation star systems, climate change, exponential increase of the human race, cosmic hazards – and for some of our politicians, trigonometry – might be said to fall into the category. Tell most people that an asteroid will hit Earth and end the human race in another week, and they will react with alarm and hysteria. Tell them that population growth, climate change, cosmic hazards and other mega-trends will destroy the human race in another 1000 years and such warnings – despite being substantiated by reams of scientific fact – will have little positive effect. Such cautionary messages will be discarded without much further thought. This type of longer-term warning becomes easy to dismiss by members of the general public because it is too far in the future and too massive in scope to be truly comprehended – to whit a 'hyper object'.

Far-sighted individuals such as Elon Musk and Stephen Hawking have talked about what it might take to colonize Mars and why it is vital to do so. Musk is developing a super rocket known as the Big Falcon Vehicle (BFV) (or Big F....ing Vehicle) that would, in theory, be able to take a significant number of astronauts to Mars. Musk has explained his vision of settling Mars and why it will take a colony of a million people to make such a colony sustainable for the longer term. Other visionaries have even more ambitious schemes. Arthur Clarke has even envisioned how 'von Neumann' self-replicating machines might, in time, be able to transform a planet the size of a Jupiter into a star so that it could generate a new solar energy source for colonies further out from the Sun.

---

[2] Meara Sharma "Why We Can't Comprehend Climate Change", Washington Post, April 8, 2018, p. B-1, B-5.

People have not grasped that we are truly on the technological verge of being able to create new types of structures in space that could save us from massive solar storms, or allow an atmosphere to form on Mars, or create a heat pipe that could pump trapped heat out into the void of outer space. We have yet to realize our potential or understand the seriousness of the threat we face from dangerous hyper-objects that threaten our longer term existence. This book is thus both a cautionary tale and an exploration of the human potential. Our challenges are technological, social, cultural and political. In the age of the Internet, the political challenges may prove the largest.

CEO of Planetary Resources, Inc., Peter Diamandis, has teamed with such partners as James Cameron in his efforts to create an asteroid mining company that will bring new riches to a global economy and would in time bring actual industrial production systems to outposts in space. Even this author, in his book *The New Gold Rush: The Riches of Space Beckon,* has talked about a trillion-dollar space economy and the unlimited resources that extend beyond the limits of the world we that we live in. Too often our political leaders have a tough time 'thinking outside the box.' Of course, this is not thinking outside of the box but thinking outside of the orb. Our "box" or "orb" on which we live needs to be seen for what it actually is. Our world is a 6-sextillion-ton orbital spacecraft that is traveling through space at 100,000 km or about 66,000 miles an hour.

There is a future potential of a vast new multi-trillion-dollar space economy and the potential of a space industry that extends beyond the gravity well we dwell within on Planet Earth. But such a possible future is generally rejected as farfetched or simply irrelevant. This is because for 99% of all humans, such things are simply not real. Such flights of fancy, science fiction, or space enthusiast speculations can be ignored as not being relevant to most people's everyday lives. The average man or woman in the street thinks in terms of things that relate to occurrences that happen a day, a week, or perhaps a year from now.

Really big and massive things that are decades in the future tend to become 'hyper objects' and discarded as being too complicated and too remote in time to be considered a topic for average people. Almost all thoughts about the very long-term future and especially large-scope subjects that are complex and difficult to understand become what Tim Morton calls a hyper-object. Such 'science fiction' does not relate to the world we live in. For most of us, our world contains mortgage payments, sports events and news, TV and movies, marriage and divorce, the birth of kids, trips to the emergency rooms of hospitals or clinics with broken ankles and other 'real life' issues and problems.

But in truth there is today a space and cyberspace revolution afoot. This revolution, that is producing new technologies, new services and new capabilities, is reshaping the world we are living in and reconstituting our economies, redefining the nature of our jobs, and disrupting the way things are done.

This technology-driven revolution that is bringing the Fourth Wave of disruptive change to the global economy also brings us the opportunity to recognize and confront the key hyper-object challenges that humanity faces today. These challenges are, in order of importance: climate change, population growth (especially in terms of megacity vulnerabilities, air and ocean pollution and resource consumption), cosmic hazards and super-automation, a global employment crisis, and jobs lost due the coming of the Cyber Revolution. There is, of course, always the joker in the deck, represented by weapons of mass destruction and potential world war involving chemical, biological and nuclear weapons.

There have always been large-scale and disruptive threats to the world economy and social and cultural revolutions that humans will seemingly always have to face, but these hyper-object threats are larger and more profound than those faced in the past. This latest Cyber Revolution that is coming to your neighborhood soon will be even bigger in scope and impact than a world war. Perhaps only the Black Death had an impact on humanity that might be considered comparable in its disruptive force, but the plague impacted tens of millions of people, while the Cyber Revolution will impact maybe as many as 10 *billion* people.

What Has Changed to Create the Cyber Revolution That Is Now Unfolding?

What has changed? The answer is, almost everything. We just don't know it yet.

The global population is rising to dangerous levels that might reach 12 billion by 2100. Worldwide vulnerabilities have risen with urbanization that has now reached 53%. This mega-trend might reach 80% urbanization by 2050. Urban infrastructure is subject to the rising possibility of terrorist attack and natural disasters. One of these mega-threats that is under-reported is the rising potential of solar storms, called coronal mass ejections (CMEs). This type of solar storm will, in coming decades, pose a much greater threat due to current shifts in Earth's magnetic field. This natural magnetic shielding currently largely protects us from solar storms, but the shields are coming down.

Scientists now tell us we have entered the Anthropocene Age, where human industrialization and population has put us on a pathway to potential global transitions that threaten the viability of our planet as a safe place for human life. These potential changes are calamitous climate change, global warming, water shortages, ocean flooding of cities and the destruction of the vital infrastructure on which over a billion people depend and many trillions of dollars are invested. The threats are increasingly clear. The key follow-on question is, can our space and cyberspace technologies and systems provide us the capabilities to reverse climate change and global warming? Can desalinization plants and water purification systems provide us the water reserves we need? Can the rise of ocean levels, the intensity of hurricanes and typhoons be reduced?

Finally, there is the additional concern of super-automation that raises a new type of specter looming over global employment. AI, smart robots, and clever algorithms will upend global employment, but with forethought, can their impact on jobs, income and fruitful human occupation be minimized? Simple extrapolation of current trend lines related to super-automation technologies indicates that almost all service jobs that exist today will be reduced. This is crucial, since 80% of the jobs of those that live in economically developed economies are in the service sector. Most employment projections are concealing or under-reporting the amount of job loss or job redefinition that will occur.

Bill Gates, who has even proposed putting a tax on smart robots, is one of the few who has envisioned the scope and scale of the Cyber Revolution. His company, Microsoft, is one of the major developers of AI systems and has one of the best overviews of how AI could impact global employment – and in a huge way. As such, he is also one of the few people who has put his finger on the size and future magnitude of this transition.[3]

The question is whether economists, technologists and political leaders anticipate the degree to which AI drives change. Will they and other leaders move quickly enough to find new and constructive answers to this massive change? Our unvarnished answer at this time is no. The leaders of the world economy are sitting on their hands and hoping that somehow it will all be okay. This is not an answer. This is, in fact, a giant cop out, and very soon we could be in deep trouble with very few answers and even fewer essential reforms. This book is a call for leadership to wake up and to start taking action.

Almost 30 years ago the author made predictions in his book *Future Talk* and *e-Sphere: The Rise of the World Wide Mind* about the rise of 'electronic immigrants' that could result in jobs being exported to other lower wage countries. As the cost and price of broadband communications and IT services plummeted, this is exactly what has happened.

Decades later, politicians bemoaned the fact that no one anticipated this problem with service jobs being shipped overseas. However, there were warnings. Super-automation and its impact on employment remains a huge problem that economists, technologists, and political leaders need to address while there is still time. Technology is creating this issue, but technology can also help to solve it. Yet we feel so helpless to solve these problems, because they are so huge, it seems best to simply ignore them.

Our business community and political leaders are not prepared for the threats that the coming Fourth Wave portends. Yes, big changes are coming, but what can the individual do about it?

No one seems well positioned to cope with climate change, overpopulation or super automation. And if the Cyber Revolution is going to create a Fourth Wave economy, where smart robots and AI algorithms take over a massive number of service jobs in the future, what could possibly be done about it?

As hinted earlier, the very technologies that constitute super-automation may also help us cope with these problems. Indeed, this might be a key part of the solution. It just might be possible that the best hope for the future would be to see how Fourth Wave systems could help address these problems.

---

[3] Geoff Colvin and Ryan Derousseau, "Bill Gates Proposes a Robot Tax" Fortune, February 22, 2017. http://fortune.com/2017/02/22/bill-gates-proposes-a-robot-tax/

The key thesis of this book is that we just might be able to use hyper-object threats such as climate change, over population and super automation as a new economic opportunity. Oddly enough the problems that confront us might also hide within them the solution to these problems.

Why Are These Large-Scale Future Threats Suddenly So Urgent?

These problems have become urgent because we live in the 'Age of Future Compression.' New space and cyberspace technologies are exploding onto the scene with unprecedented speed. The future is coming at us faster and faster. What was once science fiction is quickly being replaced by science fact and beta-testing of systems based on space and cyberspace technological reality. We are no longer just having future-oriented thoughts about space travel or solar power satellites or smart robots that in coming years will likely become as capable as humans.

IBM's Watson is here and being applied to medical diagnosis, to smart city planning, and even to filling out tax returns. We have filings at the International Telecommunication Union for large-scale low Earth orbit constellations of small communications satellites and remote sensing. These innovations and much more are already here or are on the cusp of happening. Plans for new space ventures are blossoming not only in places like Silicon Valley but also in Toulouse, France, and Star City outside of Moscow, Russia. Further, remarkable new space ventures are underway in places like ISRO headquarters in India and the Chinese National Space Agency as well as with a number of start-up Chinese space companies such as CASIC that is developing the OneSpace launcher.[4]

At the 2018 Space 2.0 Conference in San Jose a group known as Orion Span announced their Aurora One 'luxury space hotel' that they suggested might be in space and operational by 2022. This modular system is projected to start with a crew of two and with six guests – if they can find paying customers that can afford $10 million for a 12-day stay in orbit.[5] And if this seems too much like vaporware, it should be noted that Bigelow Aerospace has both its Genesis 1 and Genesis 2 spacehabs already up and flying as beta tests for their space hotel. Robert Bigelow, who heads Budget Suites hotels on Earth, is hoping to play host in Earth orbit (Fig. 1.1).

However, 'Space 2.0' driven products and services are only a part of the evolving new economy that are symptomatic of the Fourth Wave economy that is driving innovation. It is actually the cyberspace industries that are truly driving our future

---

[4] Jeffrey Lin and P.W. Singer, "Watch Out SpaceX: China's Space Start Up Industry Takes Flight," *Popular Science,* April 22, 2016. https://www.popsci.com/watch-out-spacex-chinas-space-start-up-industry-takes-flight

[5] Maureen O'Hara, "First luxury hotel in space announced", https://edition.cnn.com/travel/article/aurora-station-luxury-space-hotel/index.html

Fig. 1.1 The mock-up of the Aurora Station space hotel by Orion Span. (Graphic courtesy of Orion Span)

and the Fourth Wave forward. Just as Space 2.0 is coming out of Silicon Valley so are new AI algorithms and a myriad of new Internet-based disruptive technologies.

Ray Kurzweil, when he was made director of engineering for Google in 2012, said that AI algorithms will be "smarter" than humans as early as 2029. He has defended this seemingly over-ambitious prediction as being far from radical:

> *Today, I'm pretty much at the median of what AI experts think and the public is kind of with them.....The public has seen things like Siri* [Apple's voice recognition software that Kurzweil developed], *where you talk to a computer and a 'computer' talks back to answer all your questions. They've seen the Google self-driving cars and almost daily new innovations. Predictions of new technologies about amazing new capabilities do not seem to be so radical any more.*[6]

Greg Wyler won the Arthur C. Clarke Innovators award in 2015 and he chatted about sustainable energy. He said that the cost of photovoltaic cells was decreasing very rapidly, and that this would allow the amount of solar energy systems to reach a level of doubling as rapidly as every year in the future. He indicated that these kinds of breakthroughs in sustainability and green energy would become commonplace when the Singularity was achieved.

What will be so different about the age of the Singularity? Large-scale employment, wealth, national power and warfare will likely be totally redefined. We might well see a single Russian multi-billionaire oligarch become not only one of the world's most powerful people but one of the world's most powerful "pseudo-nations." Let us repeat that, so that it can sink in completely. In the area

---

[6]Adam Withnall "World's leading futurologist predicts computers will soon be able to flirt, learn from experience and even make jokes", *The Independent,* February 23, 2014. https://www.independent.co.uk/life-style/gadgets-and-tech/news/robots-will-be-smarter-than-us-all-by-2029-warns-ai-expert-ray-kurzweil-9147506.html

of super automation and the Singularity, a person can buy artificially intelligent soldiers, subjects and military power. We did not want to believe that when we saw a possible future in the movie *Star Wars: Attack of the Clones*. Yet, such a future could occur. Perhaps in 20 years, one could invest a few billion dollars and buy a totally dedicated clone or AI-controlled robotic army that one person or computer program could totally command.

Such a robotic army of clones, controlled by only one person or a powerful cadre, could become overnight one of the top hundred military armies in the world – if not one of the top fifty. The old equations of power based on land area, human population, or agricultural production are suddenly evaporating into the shadows of the past.

At a Comsat conference in Washington, DC, in the early 1980s, Arthur C. Clarke said that the most important invention of the twentieth century was artificial intel-ligence. At the time the HAL computer from *2001: A Space Odyssey* sprang to mind. None of us at the time, however, had a clue as to what a specter of sufficiently powerful robotic machines of the future might bring. Today we are just beginning to think about AI-enabled robots and what they might be able to do. Ultimately, they can and probably will change every dimension of modern life. We are just beginning to understand what such a future reality will truly mean for jobs, warfare and power. Again, the key question is whether these powerful new technologies can be used to address the biggest challenges that face human society.

Everyone from Isaac Asimov to Elon Musk – and even several of the *Terminator* movies – has warned us about a future dominated by AI machines. In most cases people think of something like "Skynet" or VIKI from the *I, Robot* movie when they think of possible coming robotic threats. They envision a future world where the machines take over the world in order to rule people. They don't realize that the much more likely near-term outcome could be a time in which unprincipled people with money and power create cyber-powered networks and machines to create their own centers of power and control. The result might be the rise of rogue corporations or even rogue nations with unbridled cyber-based power. It is this cyber threat from AI-enabled robotic systems that could constitute a much more real and much nearer term threat. This is not to dispute that the longer-term threat could come as well.

Ultimately the biggest near-term threat might be not finding creative ways to use smart machines and AI algorithms to address the collective 'hyper-object' threats of climate change and overpopulation. The most difficult challenge of all might ulti-mately turn out to be super-automation – in terms of unemployment and social and economic displacement in society.

There is radical and fundamental change coming. This change, if we are not care-ful, could have a devastating impact on just about everything. The nature and very definition of work, education, health care, warfare and even human survival as a species are going to be swept up in the sea-change we are calling the Fourth Wave. But this change will be driven not by clone-equipped armies. It will not be driven by families buying a servant robot to clean their houses and cook their meals. Much of the fundamental shift in the world economy will be driven by the "NewSpace" economy and all of the amazing new technologies that are generating a cornucopia of new capabilities.

Entering the Fourth Wave Economy

Alvin Toffler in his groundbreaking book *The Third Wave* explained how humans evolved for millions of years as hunter/gatherer nomads until, about 10,000 years ago in 8000 B.C., they discovered the advantages of farming and agriculture. With the planting of seeds and the cultivation of crops they could create towns and cities. There could be specialization of skills, and new professions blossomed; people could be warriors, architects, craftsmen, etc. This was the first wave in the evolution of human civilization.

With the Renaissance there was a burst of new knowledge and the rise of scientific experimentation and engineering techniques. This enabled the birth of the Industrial Age. At this time 90% of the population of the world worked on the land as farmers, miners or related labors. Over a period of some 200–300 years the employment patterns around the world changed. More and more people worked in factories to produce products and machines. Those machines included tractors, harvesters, seeders and plows that allowed the industrialization of farming. Other machines enabled the mechanization of mining. Today, in industrialized countries, about 3% of the workforce is engaged in farming or mining. The Industrial Age, or the Second Wave, lasted from around 1700 up until the middle to latter part of the twentieth century.

In the 1950s the industrialized plants that made machines and products began to be automated. These plants could increasingly produce computers, processors and smart telecommunications and information technology products that could replace the industrial worker in more and more tasks. Fewer and fewer workers were needed on assembly lines that were not only automated, but the plants could operate 24 h a day with software that could direct everything.

The Third Wave, based on a service economy, increased unemployment of industrial workers. Workers on assembly lines hit their employment peak in the 1960s. There are many workers still in industry today, but more and more their jobs are in services such as sales, accounting and various forms of planning and management. Toffler's books, *Future Shock* and the *Third Wave,* and David Bell's books about the post-industrial society, were about how the service economy was driving employment and economic growth.[7]

The First Wave lasted 10,000 years, the Second Wave 300 years, and the Third Wave gave birth to the service economy, which has now lasted some 60 years, but its days seem numbered.

The Fourth Wave will be the age of NewSpace and AI and the new technologies that they will engender. These will include 3-D printing/additive manufacturing, sustainable and renewable energy systems, self-aware machines (SAMs) and intelligent robotic systems, plus a remarkable range of Internet-based applications that disrupt established industries and services. There will also be a host of other

---

[7] David Bell, *The Coming of Post-Industrial Society: A Venture in Social Forecasting* (1973) Basic Books, New York.

developments that include 'smart' drones, reusable rocket launchers, on-orbit services and more. The Fourth Wave will be fueled by new industries and challenges that will be spawned by the trillion-dollar NewSpace industry and AI systems and will redefine what people do with their lives, as the meaning of work and productivity are reinvented.

The presidential election of Donald Trump in the United States, the passage of Brexit in the UK, the rise of Marine Le Pen's popularity in France and the shifts of political allegiances in Eastern Europe are not accidents. In a variety of ways these 'popular rebellions' are all fueled by those who see technological innovation and the impending arrival of the Fourth Wave as a bad thing rather than as a potentially good thing. They are concerned about the loss of jobs in industrial rust belts that are visual reminders of the downside of the shift of employment from the Second Wave to the Third Wave. Voters in the United States and Europe are worried about the prospect that new technologies and automation could decimate the service jobs of the Third Wave, on which so many workers depend. They are beginning to fear that the Fourth Wave economy might well erode all but high-tech jobs as super-automation advances.

Political leaders will – as they have so many times in the past – ask the stunningly uninformed question: "Well, why didn't someone warn us this was coming?" The answer is, of course, that they did, but it was preferable for them to ignore the "dying canaries" that warned us of technological unemployment, the downsides of super automation, the hazards of global overpopulation and super urbanization. Even the consequences of climate change are not fully understood by the electorate, though many sensed they were going to be a part of a diminished future.

Coal miners, assembly line workers and millions of others who are seeing their jobs disappearing are scared, angry and confused, and see no hope for the future. With trends associated with super automation, the future is clearly in a state of transition. No politicians are going to be able to bring back the lost jobs from the past. We are indeed entering a new economy that is driven by automation, artificial intelligence, big data, networking, the NewSpace systems and disruptive economic concepts that are centered on education and learning new skills.

However, these populist rabble-rousers are able to stir emotions, especially against immigrants, and exploit the angst that comes from the rapid economic transition now so prevalent in today's world. This means that the modern world is at a critical crossroads.

Several key trends are happening at the same time:

- New opportunities are opening up as a result of disruptive new technologies, such as labor-saving devices and self-aware AI, and a wide range of new industries that can speed the rate of innovation.
- Several key problems are confronting the world economy in tandem. These include: (i) rapid global population growth and even more rapid urbanization; (ii) climate change and growing global pollution; (iii) threats from terrorism (including cyber-terrorism), rogue nations with chemical, bacteriological and nuclear weapons and religious extremism; (iv) inadequate education and health

care systems; and (v) new economic and employment concerns that arise from super automation and the Singularity, which will compound job-related and economic issues associated with the Fourth Wave economy.

- Political leadership, regardless of their orientation, has been slow to recognize the complexity of the technological, environmental, economic and demographic issues that face modern society and the need for fundamental shifts to adjust to a Fourth Wave world. Instead of reform and basic adjustments to the above problems, there are emotional appeals to supporting populist agendas and in some cases extremist appeals to racism or anti-immigrant sentiment. This will only hinder the possibilities of rational response to current challenges that the Fourth Wave constitutes.

A Quick Tour of the Cyber Revolution

In the chapters that follow we will present the challenges of the Fourth Wave and possible strategies, using proactive planning (or futuring) to help us survive a time of great turmoil and transition – perhaps greater than humankind has faced to date.

We will explore hyper-object threats in more depth. This also entails examining in greater detail the key dimensions of the threats related to overpopulation and the growth of megacities, climate change and pollution-related problems, and how new capabilities can be applied to the challenges posed by the mega-trends that could threaten our future. We will first explore the nature of the major hyper-object challenges to be faced in the next few decades and then possible NewSpace and cyberspace responses that can help respond to these large-scale problems. This will examine how we might have to adapt our economic and technological practices and as well as our industrial production, supply, and market-based systems to meet the challenges.

As a first step we need to better understand the challenges posed by climate change, surging demographics, super-urbanization, natural disasters and especially cosmic hazards, including those that could become more deadly in threatening vital infrastructure as well as the coming era of super-automation and the Singularity. All of these forces will reshape our world in powerful ways. Once we grasp the full scope of the threat, we can begin to find ways to marshal Fourth Wave systems and technologies to address these so-called hyper-object challenges – politically, economically, environmentally, technically, socially and culturally.

We will explore how the expanding new space and cyberspace industries could help us address these various twenty-first century problems that we have characterized as the Cyber Revolution. These growth industries have many applications that go beyond coping with major challenges. The prime focus of this chapter, however, is to examine how these industry innovations can help to address the problems of overpopulation, overcrowded cities, and climate change and accommodate the changes that come from super-automation and the Fourth Wave economy.

Exploring the potential of smart cities and proactive planning is also covered, as a key response mechanism. This means we will explore techniques developed and tested in modern urban planning and the creation of smart cities, to see how this can help. We will begin to examine the need for and opportunities presented by greater cooperative economic and political actions in the smart city era. This is critical, since 80% of the world's population will live in an urban environment. Ironically the growth of the Internet and global social networking may form a valuable tool to support proactive planning, but these same capabilities also may also constitute a major barrier to the use of these techniques within modern democracies.

What type of education and health care systems are best suited to the needs of society in the decades ahead? As the many diverse demands of the Cyber Revolution become clearer and the challenges more demanding, we need to find way of coping with a world community under pressure to reform.

What is the meaning and the impact of the so-called personal communication revolution and nearly universal broadband networking in the age of the Cyber Revolution? There will be particular challenges related to living and thriving in the world of the Internet that will ultimately become the Internet of Everything. This world of broadband services and omni-present social networking will create its own unique set of challenges, which will include such issues as the lack of personal and societal privacy, the invasive nature of ubiquitous automation and the need to cope with constantly changing technological systems and software that requires mental agility to adjust to its latest iteration. In this world of universal Internet and constantly adapting software humans will strive to keep pace with ever faster and demanding smart machines that never rest. On top of it all will be the need to cope with information overload.

Finally, we will provide an overall analysis of the many twenty-first century challenges that will beset our spaceship planet in the Cyber Revolution. By the end of the century virtually every aspect of life as we know it today will have changed. It is essential that we change the ways we cope with climate change, the ways we seek to control urbanization and population growth, the ways that we plan our cities, the ways that we provide education and health care, the way that we share knowledge and use broadband communications and the way we live and earn an income. This will all change as we share our planet with smart robots and as the Singularity takes root and alters life on our planet. Some people feel that this change will be scary and disruptive, and the 'progress' that is coming is really not progress. The future, however, is a one-way gate. What humans may become will be revealed by the technological future, and we must be ready and adapt to change. We must find ways to marshal the force and intelligence of high tech industry to save our planet from overcrowding, climate change and super-automation. We must not only survive the Cyber Revolution but advance the human race to the next stage in its evolution. The twenty-first century represents a narrow and potentially perilous pathway to the future. Let us all hope we can succeed in creating a better future.

New Directions

The twenty-first century is a time of great transitions. We are going to see massive changes in the environment, global politics, urban planning, and in economics and business systems. We will also see massive changes in our demographics, educational and health programs, and in how we use and invent new technologies; this will lead to all sorts of institutional reform. The overall result will be massive changes in how we live, work, procreate and move to becoming a multi-planetary species.

The challenges are everywhere. We need to cope with the almost explosive transition from the Third Wave to the Fourth Wave economy that will redefine the nature of work. If anyone thinks that today's world is in a state of excessive change, just wait for the next fifth. The biggest dominoes are yet to fall.

In the 1950s and 1960s the U. S. Agency for International Development (AID) brought to El Salvador a distance education program that was aimed at schooling the rural population, and it worked well to educate rural youth. The problem was that when the newly educated youths graduated from these new educational programs there were no jobs. The result was revolution. Today Silicon Valley is bringing the world wonderful new technology that includes big data, artificial intelligence, expert systems, smart robots, wonder drugs and broadband data networking. This is super-automation. These smart knowledge products and services have brought about labor-saving systems and a population that lives longer and with many increasing social benefits. But the purveyors of new cyberspace products and services have not brought to the U. S. electorate what they wanted. This was an array of meaningful and well-paying jobs.

In some ways this mix of super-automation, the challenge of climate change, overpopulation, rapid urbanization, longer life through improved medical care and life-long education coupled with a shrinking number of meaningful and well-paying jobs is a heady mix that is explosive – even dangerous. To many, including this author, the future is fraught with some considerable peril. We have had a foretaste of this revolutionary cauldron of frustration in the U. S. 2016 Presidential election. Both Donald Trump on the right and Bernie Sanders on the left appealed to sectors of the U. S. electorate that wanted change. Many of the electorate – on the left and on the right – were frustrated with the status quo. Voters in rural America perceived that their wants and needs have been overlooked and that technology or immigrants had taken their jobs. They voted for whom they thought might serve as a change agent to address their grievances. Voters on the left wanted to see more social services, free college education and someone that would fight for them.

In the end, virtually no one got what they wanted or expected in the election. Nor was the United States unique. The Brexit vote in the United Kingdom was clearly led by a protest vote against the Common Market and the European Union. The vote in Germany that weighed heavily against Angela Merkel was a protest vote against unwanted immigrants.

The ingredients of change are everywhere. New modes of manufacturing and production, especially additive manufacturing and 3D printing, are being produced by the cyberspace industries. There are the NewSpace entrepreneurs and the rise of what is called by some Space 2.0. In this arena, we see the Silicon Valley innovators invading the turf of the traditional aerospace industries. Everywhere innovation is redefining how and why we do things. Everywhere one looks we see cyber-savvy people re-inventing society via an impressive network of new IT and smart services. Add new software, and one can re-invent the world of taxi cab services (i.e., Uber and Lyft). Add software, and Air B&B re-invents the hotel world. Add software, and the retail industry can be upended (i.e., Amazon and Alibaba). And the revolution is just beginning. Software and cyber smarts will in time significantly reinvent education, health care, political campaigns, vehicle design and manufacturing, and perhaps even government and governmental services.

The net result is the increasing eroding of jobs as we once knew them and disruptive technologies expanding across the world. The trends lines as a result are askew. We see populations still expanding, automation expanding, underemployment or unemployment increasing, the rate of increase for climate change accelerating, and political cohesion increasingly strained. We see massive changes that are contending against other. Expanding population is really the worst thing to contemplate as we look at automation, the Singularity and climate change. Toss into this circumstance illegal immigration and streams of refugees from war-torn and gang-infested countries, and one can see a cocktail for social unrest and increasing political strife.

All the while we see new and brilliant conceived technology being churned out of Silicon Valley. Of course, Silicon Valley is no longer just in southern California. It is also in southern India, in research labs and universities in China and has spread across the world. In Cape Town, South Africa, there is a high-tech company called New Space that is manufacturing key components for smart cube satellites the size of a tennis ball canister. In the age of broadband Internet connections advanced technology can be everywhere.

Disruptive technologies and clever new products and services can bring change and innovation, but most societies can become politically and economically unstable if there is too much change and too many people feel that the change that is occurring is hurting rather than helping them.

What we may find is that democratically governed countries are more fragile than we had thought to be the case. Democratic governments have been relatively stable, and economic growth has largely been sustained since the time of the end of World War II. But a new age of change seems to be coming.

Software developers and artificial intelligence (AI) programmers are busily at work to replace and largely make redundant what accountants, property appraisers, bankers, insurance brokers, and a host of other professional service providers now do for a living. Uber may have generated new jobs for part-time drivers, but its ultimate business model is based on driverless taxis (See Fig. 1.2).

Airbnbs may have provided a new source of income for some households, but its business model is aimed at undermining the traditional hotel business. Google's and Telsa' driverless cars and vehicles will not only impact taxicab drivers but will

Fig. 1.2 Driverless cars – benefit, bane, or both? (Credit: Internet commons)

eventually take its toll on truck drivers, bus drivers and more. The Space 2.0 industry is likely to create lower cost remote-sensing systems that can replace a wide range of jobs in law enforcement, urban planning, farming, and environmental protection. Automated space systems or drones will likely automate many communications, information or delivery services now carried out by humans.

This is not to attack companies involved in cyberspace services but rather to suggest that they should recognize that their business models are short-sighted. The first reform in their thinking must be to see how their products and services can ease the conversion from the Third Wave economy to the Fourth Wave economy. In doing this analysis they must see how they can also begin to maximize gains for a global society beset by large-scale challenges. As noted earlier Bill Gates has suggested a tax on robots to aid this transition. Others have focused on providing a living wage for all people in a society. Yet others have focused on constraints on population growth and guarantees of education and health services within a highly automated society.

The bottom line is that there are truly a number of threats ahead that must be faced. There are at least three challenges or threats to human society that are at the core of the Cyber Revolution, which must be addressed and solved by the end of the twenty-first century. These three challenges, mentioned earlier, are so large that they can accurately be described as the hyper-object threats:

- **Overpopulation.** This includes concerns for life extension that is increasing the ranks of the retired workers for decades longer than in the past, and requirements of the elderly for medical care and pensioner benefits. There is the related problem of the world crowding more and more people into megacities as urbanization rises from 53% to perhaps 80% in the next 50 years. Ultimately it is the requirement for fewer people when the global population is still soaring upward. One metric that is probably on the optimistic side is that for every 50% increase in human population the chances of survival for the human race is cut in half.

- **Environmental Damage and Climate Change.** There are still increasing amounts of air and ocean pollution, release of greenhouse gasses into the environment, thawing of the icecaps and of the Siberian frozen peat fields, the release of methane from massive mines and herds of cattle, the continued expansion and operation of coal-fired energy plants and so on, all accelerating climate change. With the rise of the global population from 7.5 billion to as much as 12 billion, there will be more consumption and more pollution. Cities that have pledged to move toward zero-carbon footprints are to be commended, but if population growth is not curtailed the problem will continue to grow.
- **A Disruptive Transition to a Fourth Wave Economy.** The key to making a non-catastrophic transition to a new economy where super-automation, smart robots and ultimately the Singularity reduce the labor requirements of humanity cannot just occur willy-nilly. A global population of 4–5 billion provides much more genetic diversity than is needed to maintain the gene pool. We need to consider how to use our smart technology, services, tax policies and political leadership to match global population with the need to stabilize climate change trends and to match labor force needs with available workers.

These are intellectual, technological, sociological, cultural, environmental, and economic challenges where the wizards of Silicon Valley and leading research centers around the world can potentially provide new and innovative answers. There is even sufficient brain-power in these locations that they can likely create new products, services, tax codes, tariff structures and other incentives and penalties that could help attack these interrelated problems.

Unless these bright, creative minds can come up with new and viable ways forward, the world's economic, political and environmental peace will be massively disturbed, if not destroyed. We not only need to stop Earth's heating up and stop the build-up of carbon-based greenhouse gases, but we need to reverse these trends and move toward a shrinking global population. It would be wonderful to find a way to create profitable new industries to move us toward a solution.

And while we are in the mood to ask for miracles, it is probably reasonable to go whole hog and ask for more, that is, holistic solutions to the following problems that are being examined in this book, such as (i) employing smart city planning techniques to cope with the problem of extreme urbanization and the growth of more and more megacities; (ii) finding new, productive and better approaches to health care and education; (iii) creating new IT systems, broadband Internet and electronic and space systems that provide for more than just higher rates of throughput or information overload to a world already overloaded with too much information, too many people, too much energy and resource consumption and shrinking opportunities for meaningful work. The need is to start finding ways to deploy new technology to address the problems rather than to simply increase profits. We have already been there and done that. Now, in the time of the Cyber Revolution, it is time for something new.

Millennials, at least in the economically developed world, can expect to live to be a hundred or more. Yet in a time of expanded life cycles they are also potentially facing shrinking opportunities for full and effective employment. Today it is coal miners and factory workers in the Rust Belt of America looking for work. Tomorrow it will be millions of college educated younger people. The parallels to what happened in Central America in the 1950s and 1960s should not be overlooked.

The Cyber Revolution and the coming Fourth Wave economy are first and foremost a warning. The future is not going to be like the past. We are facing a period of radical and unprecedented change. This is not a call for a new and modified diet. It is not a call to be more conscientious about recycling or to buy a Prius or to learn a new trade or skill in case super-automation happens to take away your current job. No – this is big time, fundamental change. It is change we are in no way prepared for, and younger people are set to be the losers – big time indeed.

This book is seriously suggesting that we are facing major changes shaped by such forces as super-automation that will lead us to 'the Singularity' around 2030 and the creation of cyberspace-related industries coupled with systemic loss of traditional jobs. There will be a continued rise of big data and cybercrime, global overpopulation, and climate change. These disruptive forces and others will combine to force humans to change – on an epic scale! It will alter our lives, our jobs, our economic systems, our family planning and practically everything else. We will cope, adapt and in time – hopefully – prosper. We will make this happen for one very important reason. That reason is survival.

The coming Fourth Wave economy will hugely impact our political systems, restructure capitalism, impact our religious beliefs and restructure how we introduce new technology into our industrial systems. And amid all these concerns and issues that will raise civilian threats and issues related to a new global economic and social system there are several hyper-objects that lurk as ominous challenges, largely hidden from the view of not only the general public but many policy makers – with a head-in-the-sand view of the future – as well.

But always there is hope. This book is not about desolation and despair. It is ultimately about hope. It seeks to chart a new pathway to tomorrow that allows us to cope with the many daunting challenges that pose very large challenges to humanity's future.

Chapter 2
We Need Better Planet Management Skills

*On a planet of nearly eight billion people with billions more on
the way, natural limits simply don't mean much.*

Erle Ellis, author of *Anthropocene: A Very Short Introduction*

The first thing we should do is to start saying to ourselves and our children: "We live on a planet that is a spaceship with limited supplies." The human passengers have to live off the supplies of our spaceship has grown too large and too rapidly. There were 800 million people in 1800, 1.8 billion in 1900, and 6.5 billion in 2000. There will be perhaps as many as 12 billion of us in 2100. We humans are big eaters and garbage producers. We are starting to be in serious trouble as result of all the resources we consume. An even larger problem is all the waste that we are producing in the form of greenhouse gases that we cannot easily offload into space.

Our spaceship travels at amazing speeds of about 100,000 km/h as it circuits the Sun. Our large spaceship actually travels about a billion km a year in its orbit and absorbs enormous energy from the Sun in the process. Despite the size of our spaceship planet, it is still limited in its resources. Earth is destined to sail through a cosmic wilderness that possesses a number of hazards for humanity. Earth will last for billions of years to come, but our species chances of survival for that long against solar storms, potentially hazardous asteroids, human-produced weapons of mass destruction, or virulent disease and pandemics – not so much.

If our planet were as large as Jupiter, for instance, its gravitational power would be just starting to be sufficient to overcome the strength of the nuclear forces trapped inside the planet's atoms. Such a giant planet would fuse atoms together and thus begin to morph from being a planet and to becoming a star.

Scientists have figured out that if you start with the mass of an atom and then increase from this atomic scale with successively larger spheres that grow 10 times bigger, then 100 times bigger, then a 1000 times bigger – and you keep doing this 36 times – something key finally happens. The mass has now grown hugely in the size and would be somewhat larger than Jupiter. At this scale the force of gravity is stronger than the strong nuclear force, and gravitational 'fusion' creates a star.

© Springer Nature Switzerland AG 2019
J. N. Pelton, *Preparing for the Next Cyber Revolution*,
https://doi.org/10.1007/978-3-030-02137-5_2

There are key 'what if' questions that come from this knowledge. What if the gravitational force were different? What if you only needed to expand the amount of mass 10^{34} times rather than 10^{36} times? Would the universe and our Earth be different? The answer is yes, and in a very big time way.

In this alternative universe stars would start to form with planets about the size of Earth. If there were to be intelligent life forms they would have to be the size of somewhere between viruses and smallish insects, because otherwise they would be crushed to death by gravity. These life forms would necessarily live on smallish planets perhaps the size of the Moon. Humans would be too large to live. They would be crushed by an overpowering gravity. The relationship between the strong nuclear force and gravitational force that is 10^{36} times weaker is known by scientists as "N," and this might be the most important design factor in a very complex universe.

Our planet, despite all of its advantages in terms of temperature, mass and natural magnetic shielding from the solar winds, is still limited to a finite amount of resources. Its total mass remains fairly constant except for stardust that settles into Earth's gravity well and thus does add tons of mass every week. If we consume too much of the planet's resources or despoil the biosphere in which we live with too much industrial activity, we could be in trouble. If we trap too many greenhouse gases in Earth's atmosphere or over-pollute the oceans, it could spell big trouble indeed.

Our planet, Terra, is thus a most valuable resource. It has a recyclable supply of water and sustainable atmospheric and oceanic systems that are capable of supporting many types of plants and animals. And we should not forget that this flora and fauna are constantly evolving. To date, there have been five mass extinction events that have combined to destroy some 99% of all the species that have ever existed on our planet. Sustaining a species for long periods of time, say 100 million years, is a tough challenge. Homo sapiens have been around only for about 4–4.5 million years. This makes us a baby species on our planet. In truth, we should be a lot less prideful in thinking we are so very special. It is not at all clear that our species represents the end and ultimate outcome of an evolutionary process. Insects have lived much longer. They have a total mass that outweighs humans by a significant margin, and their chances of survival for the longer term seem to be much higher. Intelligence and long-term sustainability within a planetary biosphere may even be contra-indicators in terms of the long-term survival of a species.

Few children, and very few adults for that matter, know these basic facts about the planet they live on, or how small the margin of error is for survival on the time scale of an eon. Some believe that the projected mass extinction event we earlier referred to as the Anthropocene Extinction could wipe out humanity and many other species within a century.

Few humans, and indeed only a small number of scientists, know the value of N is 10^{36}. N is actually a very magical number, with huge cosmic implications This magical number represents a proportionate ratio in the strength of the strong nuclear force and the gravitational force. It is the number that divides the cosmos into stars (masses exceeding N) and planets (masses whose sizes are below N). N might be referred to as an inside joke among astrophysicists as the 'GUTS' of existence and life within the cosmos.

N is the huge number that reveals that when a planet reaches the size of a Jupiter (or 10^{36} times the size of an atom) it is on the verge of becoming a sun because its gravity is strong enough to start a nuclear fusion reaction and overcome the strong nuclear force within the atoms. To visualize this, think of a sphere around an atom. Then think of another sphere whose radius is 10 times larger and keep going a total of 36 times.

Such knowledge should let our technologists and leaders see Earth and humanity as representing a rare occurrence in the universe. We humans should learn to take better care of our special planet if we want our species to survive and thrive.

The Three Hyper-Objects That Imperil Our Future

The likely progression in our knowledge in just a few decades should be quite staggering. The prospects for new technological and scientific knowledge that come from R&D, super-automation, AI and ultimately the Singularity are enormous. There will likely be exponential increases in what might be done with regard to population control and family planning and related health care, with regard to climate change and with regard to employment and economics and tax policies. We need to understand the most effective and efficient technological and scientific responses to these challenges. In short, we need to assess and use this new knowledge effectively, dispassionately and – in a word – wisely.

Unfortunately, twenty-first century political and economic systems seem disproportionately weak in comparison to our technological and scientific systems. Plato, in his book *The Republic,* noted this problem in human political society two millennia ago, so the issue is hardly new. Two thousand years later the problem remains.[1]

Today, economic and political leaders define what we do with scientific and technical knowledge. The result is to produce more and more products and services to generate profit or to increase political or military power. For a few centuries this system has led to modernization and unparalleled riches for certain nations. These trends to make agriculture and industries more efficient has had a counterproductive result. It has, in part, led to what now seems unsustainable human population growth (The age of industrialization spurred human population growth as never before. This growth resulted in a spurt from 800 million people in 1700 to 7.5 billion today and perhaps a peak of 12 billion in 2100).

This rise in population and affluence, at least for some, has not brought about political stability, nor truly efficient economic systems nor world peace. What we are lacking are: (i) ways to curtail runaway population growth; (ii) truly effective ways to cope with climate change; and (iii) a political and economic system that can

---

[1] Plato, *The Republic.*

create a sustainable world. We are far short of a strategy that can sustain life and create a viable longer-term global system that can bring affluence, knowledge and sustainability to a 'global tribe of humanity.' We are far short of a united front to a visionary future. And we will need a new unity in order to be optimistic for the future. We are short of such future-oriented goals that could create a multi-planetary society that could also protect our spaceship planet from cosmic destructive forces or identify a strategy to avoid a mass extinction event.

The twenty-first century is one in which the adults in the spaceship should help to redefine our priorities, our global vision and reform our economic and political systems to move in new directions. In 2001, when this author accepted the Arthur C. Clarke Lifetime Achievement Award, my talk warned of the three challenges to humanity's longer term survival. It noted that the threats that were perched on the cusps of the twenty-first century were large, worrisome and incredibly difficult to solve. Today, nearly two decades later, we are now much closer to the precipice, and human political and economic policies have served to bring us much closer to the precipice.

In short, scientific, technological, economic and political leaders should be pointed toward trying to solve problems greater and more meaningful than the often superficial agendas we now seem to have. There are much more important things to be accomplished than how to expand a 100-channel cable television network to 400 channels or streaming more football games to broader band cell phones, or how to post more videos on Snapchat, or designing a more stimulating, multi-colored, sensitive and form-fitting condom. Indeed, given our population expansion crises, we should probably be designing more uncomfortable and durable condoms.

In short, we should be examining how we can use our space and cyberspace technology to create a more sustainable world for the longer-term future. We need to look to using our best technological systems and capabilities to cope with the very difficult to comprehend hyper-object problems that confront humans in the twenty-first century.

In particular, we need to start thinking now about how to use technology, regulation, a system of laws, and a less flawed form of capitalism to address mega problems that transcend national problems and the petty day-to-day politics and issues that consume so much of our time and energy, including efforts to earn a living, pay the rent, and amass something that we call wealth. It is the great luck of astronauts to be able to orbit Earth and see the great reach of our Solar System in which our planet exists, and which is what Carl Sagan, the astrophysicist who narrated the popular television show *Cosmos,* so accurately called in his book by the name *The Pale Blue Dot.*[2]

We need to think about how to use the power of Space 2.0 and Cyberspace 2.0 not only to grow the world economy but also about how to harness that power to sustain the human race. This means likely goals such as shrinking human population, reversing climate change, finding new ways to employ humans in the world of the

---

[2] Carl Sagan, *Pale Blue Dot: A Vision of the Human Future in Space* (1994), Random House/Ballantine Books, N.Y.

Singularity, and so on. Without a viable world to live in there can be no wealth, no economic growth, not anything to strive for – no future.

The threats posed by our times makes the twenty-first century the most dangerous and important time in human history. Four million years of human evolution will pass through a "crucible of challenge" that is the next 50 years of human existence. If we can change, innovate and adapt, then a diversity of life forms can survive on our planet – including humans. Without adapting and innovating in the ways outlined in the chapters that follow we have a high probability of perishing as a failed evolutionary project that placed growth and accumulation of material wealth above all other goals. But if we continue to ignore the demands of sustainability – we can fail, as have 99% of all species that have existed on Planet Earth up until this time.

There have been five mass extinction events in the last 400 million odd years, and there is already evidence that the sixth mass extinction event – driven by human existence – is now afoot. This is the Anthropocene mass extinction event, as mentioned in Chap. 1. But there are many ways that this calamitous event can still be avoided. Sustainability can be achieved. A new revolution in global markets, innovation and human employment can be achieved. This can be driven by Space 2.0 and Cybernetics 2.0 innovations. Technological innovation and a new type of human prosperity can be a part of the solution. Yet hard choices remain, such as those related to abandoning 'throughput consumption,' birth control, recycling, renewable energy, and new incentives and pricing models in economic systems.

For the last few centuries, human economic systems, especially those in the dominant countries of the world, and their political and economic leaders have defined the prime goal of global society as being growth of population and industrial systems. The time to change these objectives and reform our economic and political systems is now. This challenge particularly falls to scientific and technological leadership and millennials who seem to be most adept in understanding why we need change and new direction.

Taking on the Hyper-Object Challenges to Save the Human Race

Over the next 40 years that prime purpose must be redefined as survival of the human species and taking on the hyper-object challenges that threaten our longer-term sustainability. Further, our current economic and political systems limit the longer-term aspirational goals that could allow our space and cyberspace technology to create entirely new forms of systems that might protect Earth from cosmic hazards, create colonies to sustain life elsewhere in the Solar System and create an economy that extends its reach across the Solar System and ultimately beyond. For human enterprise to reach a truly visionary future, it must veer from the limits of its past. Humanity must graduate from its colonial and tribal past to a visionary 'Terran view' of what humans could be on a cosmic plane. We need to fabricate megastructures in space to defend Earth against cosmic hazards and indeed harness the potential of outer space that hold riches that extend well beyond the limits of a single six sextillion ton space ship planet.

Why do we need to make some of these changes now?

This is a fair question, and the rest of this book seeks to address why we need to seek new solutions to these problems and do so with even greater urgency. This is simply because the longer we wait, the higher the costs of remedial programs. And if we wait too long, the whole future of life on Earth could be at risk. Of course, for the pessimists in our midst, they might say we have already left it too late. So, hopefully, there are some optimists left that will read on.

The whole point of this book is to convince you that the challenges are real. Further, longer-term human survival is tied to taking on at least the following three challenges. These challenges are:

(i) **A too large human population.** Our growing human population consumes too many resources, pollutes our planet, creates a too large work force for a Fourth Wave economy and is dangerous from a tribal warfare perspective;

(ii) **Climate Change.** The dangerous implications here are manifold. They include global warming, sea-level rise, the melting of the icecap into sea water that will not re-freeze, devastating droughts, deadly hurricanes and typhoons, and more. These are all potentially devastating in the decades to come. In the longer term too much trapped greenhouse gases in the atmosphere spell out mass suicide for all life on Earth. Economically, the longer we take to cope seriously with climate change the higher the cost.

(iii) **Conversion to a Fourth Wave Economy**. This challenge comes from the fact that technological innovation is essentially a one-way gate. The advent of super-automation, smart robots, AI, and the Singularity tends to automate all forms of farming, industrial manufacturing and services. These innovations add ease to the human condition but ultimately render humans, as a work force, largely passé. We need to think ahead as to what this means for education and health care, employment, cultural and artistic achievement and the need for population control. In this world, where smart machines do all the work, does everyone just automatically get a living wage salary and what do humans do with their lives?

Obviously these three challenges are interrelated, and all three have complicated cultural, economic, ethical, ethnic, religious, political and societal implications.

All three of these challenges are very hard to comprehend. This is because of the scale of the problems they present to the man or woman in the street. These challenges are all complicated and are very difficult for any one individual to take on. These are what might be called planetary problems. Unfortunately, as such, they become easy to dismiss. They thus become someone else's problem. This is simply no longer acceptable.

The way forward must be based on renewable resources and will ultimately involve many other macro-changes to human society as we know it today. These challenges are just too large and span too wide an arc of time for easy resolution. Yet true macro-changes in our political, economic, cultural and social-ethnic-religious systems will be the only way to respond to twenty-first century challenges.

All three transitions in terms of population controls (incentives and disincentives), climate change (regulations, economic controls and tax policies), and Fourth Wave regulatory and legislative policies at the national and global level, represent a very hard transition to make. These transitions and why we need to make them will be very hard to comprehend because of their vast dimensions in time and space. The current political world is seemingly based on anti-immigration policies, ethnic and religious hostilities and a lack of global recognition that we are truly spaceship Earth in need of a global systemic process. Humanity and our world versus today's challenges could unlock the potential of a set of policies that could create an economy where new space and cyberspace technologies could then unlock a whole set of new inspirational goals and economic opportunities for the human species as it becomes 'homo electronicus.'

Focusing in on the Major Challenges We Face

Over Population

For centuries growth of the human population has been a good thing. Such population growth has provided more workers, more consumers of goods, and a mode of economic growth. But in the twenty-first century in the Anthropocene age of super-automation and climate change a growing population of humanity is now a liability. We are facing an age of overpopulation, over consumption of natural resources, rising temperature levels, spreading pollution, runaway energy demands, technological unemployment, escalating health and education costs, information overload, loss of privacy, local, national and international threats to security, a broadband global Internet that holds both opportunities as well as clear vulnerabilities, as personal information collected via social media is being 'weaponized' and 'monetized' against consumers. We see a pattern of runaway greenhouse gas emissions and climate change that could threaten the very existence of humanity a few generations hence. During the transition to the Fourth Wave economy we will face unique new challenges, and many of the challenges will combine in such a way that their solution becomes increasingly difficult. The longer we wait, though, the higher the costs of recovery will become.

In short, it is imperative that we soon find new solutions to unique twenty-first century problems. We must find out how to provide employment for people when technology-driven "productivity gains" are increasingly tied to surrendering more and more jobs to super-automation. We must find out how to change from today's world of capitalist advancement, where we are taught that the prime values (i.e., the prime goals) are always growth and wealth accumulation. Actually survival – not growth – needs to become our prime goal.

Climate Change

Climate change is hard to see because it is so very gradual until it reaches a point of no return. The rise of sea levels comes millimeters at a time. The destruction of the ozone layer comes at a molecule at a time. Genetically damaging radiation to humans and other species affects one gene at a time. The lack of water impacts a city or a region but not the whole world all at once.

In some places on Earth a true water crisis already looms, and a projected "zero hour" may come in 2019. But humans are good at denial. It is easy to always think "Oh, that is an awful problem, but fortunately it cannot happen here."

But it *is* happening, here and everywhere. The dimensions of this challenge are so varied, interrelated and difficult. They include the release of greenhouse gases, the rise of seawater levels, the flooding of vital infrastructure, the melting of the peat fields in Siberia and the polar icecaps, the increasing number of violent storms, the increasing cost of insurance coverage, and the need to provide food, water and more industrial products and services for an ever growing number of humans in a world that is also facing unemployment as the rate of super-automation increases. Steps forward such as the Paris World Climate Accord are just baby steps forward in relation to what needs to be done. We keep thinking in terms of what political forms of agreement are possible, rather in the hard realities of what needs to be done to survive.

Converting to a Fourth Wave Economy

Each transition that has occurred in the world economy has come faster and faster and has involved more and more people. The first wave was the shift from humans being hunter-gatherers to a farming society built around towns and cities. The second wave was the conversion from being largely a farming economy to being largely an industrial society. The third wave was the conversion of major employment being centered on industry and manufacturing to a service economy. The fourth wave is the post-services economy where super-automation, AI, smart robots, and eventually 'the Singularity' leads to a post-work world where economic systems and human population growth are not now in synch with the coming global economy and its labor needs.

How to Proceed?

Today we have made much of life on Earth a "zero sum game." This is a game in which if someone gains then someone else must lose, so the net result adds up to zero. In this global zero sum game there are now two adversaries. One contestant or goal is that of personal growth and riches. The other goal unfortunately turns out to be survival in a resource-depleted and polluted world drained of energy and choking with greenhouse gases. Standing in the way of growth and riches is now what ecologists call a "sustainable world." This is a stable world that can last for eons as a viable habitat for advanced biota, including humans. If we are clever we can live well and, rather importantly, also survive. If we are not clever something rather unfortunate can happen – we all die.

In a world where super-automation can insure wealth, it can also foment unemployment, over consumption and runaway climate change as well. The problem is not the technology *per se*. The problem is economic systems that do not put a high value on survival, a stable climate, and sustainability. Today's *laissez faire* capitalism and sustainability are locked in a dangerous battle that need not be a fight to the death. We can become smart enough to redesign our economic systems to allow people to live well and also survive. We need not all become communists or live in caves. We can simply restructure capitalism to meet twenty-first century needs.

We predict that the combined forces of climate change, super-automation, global networking, super-urbanization, new forms of education and health care and new approaches to security will result in changes everywhere within the global economy. These changes will come in all countries of the world. The changes will come everywhere – in towns, cities and rural areas. New technologies, new regulations and legal systems and especially restructured forms of capitalism will unleash a Fourth Wave that will allow the survival of a human civilization that has taken some four million years to evolve to date but could be irreparably destroyed in a mere four decades from now. The choice is ours.

The danger is real. We were warned nearly a half century ago by people like Paul Ehrlich in *The Population Bomb* and Harrison Brown in *The Challenge of Man's Future* that the dangers were mounting. What Ehrlich and Brown were warning us about was not that we were running out of food, like a modern day Malthus. They, and more lately James Lovelock, have been telling us that we humans were, in dozens of ways, creating a non-sustainable ecological system that would not be able to continue current levels of exponential growth – in terms of natural resources, runaway pollution and especially more and more people.[3]

We predict that these forces of change can and will modify the goals and objectives of today's growth-oriented capitalist economies. These forces of change can and will disrupt our political systems. The transition forces are strong enough that they could also result in major disruptions to global peace. In the midst of this turmoil, global population growth will become more and more controversial. Soaring populations in some of the world's largest countries could become as controversial as making and stockpiling nuclear weapons.

The fears will mount as the dangers of continued population growth become better understood. Recent studies undertaken in the United States (in Oregon) and the UK (London School of Economics) project that a typical baby living a typical

---

[3] Paul Ehrlich, *The Population Bomb*, (1971) Ballatine, New York. Also see Harrison Brown, *The Challenge of Man's Future* (1964) Viking Press, New York.

Fig. 2.1 Polar bears are just one of many large mammals that are endangered today. (Credit: IStock photos)

lifespan in America will generate an estimated 1644 tons of carbon dioxide. The only good news in this study is that an American baby's pollution potential outstrips the rest of the world. A Chinese baby would produce an estimated 350 tons of carbon dioxide and a baby in Bangladesh would only produce about 20 tons of carbon dioxide.[4] The bad news is that total world population, now at around 7.5 billion, is projected to reach 12 billion by the end of the twenty-first century. This projection takes into account that some countries such as Japan and Italy have already reached zero population growth, but in many other countries populations continue to swell.

As Roger Martin, chairman of the British-based Optimum Population Trust group, has said: "There is no possibility of drastically reducing total carbon emissions, while at the same time paying no attention to the drastic increase in the number of carbon emitters."[5] The problem, unfortunately, is even worse than climate-change activists might think. We are entering an age where more babies not only mean more pollution and accelerated climate change, but more and more of these babies are likely to be unemployed. Of course, our continuing population explosion is not only endangering the future of human civilization but essentially all large mammals are endangered as well (see Fig. 2.1).

---

[4] David Fahrenthold, "When It Comes to Pollution, Less (Kids) May Be More," *Washington Post,* September 15, 2009, P. A3.

[5] *Ibid.*

While lots of parents around the planet are producing more and more new progeny, scientists and engineers are designing and developing smarter and smarter machines that can replace more and more jobs. If one takes the example of the high-tech field of satellite communications and look at its economic figures for 2009 it shows a pattern that can be seen in many other areas as well. Earth station sales were up, communications satellite sales were up, rocket launch sales were up, and communications satellites services around the world were up as well. But what went down was employment in all of these sectors. Automation was making these industries more efficient, more productive and more profitable, but jobs in these industries declined.[6] More babies in the twenty-first century mean not only more greenhouse gas emissions but more and more adults headed for the soup lines when they find there are no jobs to be had.

Super-automation that will inevitably lead to technological unemployment under our present forms of *laissez faire* capitalism will force us toward a new understanding of work and economic compensation within the next two to three decades. We will be forced – in a very brief period of time, in comparison to, say, the Industrial Revolution – to cope with very powerful disruptive forces of change. These drivers will include artificial intelligence, self-aware machines, globally interactive social networking systems, super-automation, climate change and more. The situation will be complicated by other social, religious, economic and political forces. Here we must consider the implications of globalization, fundamentalist opposition to Westernization, etc., that come to us from the "outside." On the "inside" we will have to cope with the escalating costs of education, health care and retirement benefits for a global population that lives longer and puts more and more stress on a worldwide economic system. Although there is much that is controversial about Kurzweil's assessment of the Singularity and how much it will help us – and, more fundamentally, how little it could hurt us – he is right on point about its potential to "rupture the fabric of human history."[7]

Just over 20 years ago predictions about telecommuting and electronic immigration made it to the top of the World Future Society's projections of major changes for the future, but these forecasts were largely ignored. About 7 years ago people began to grumble about companies moving their offices, plants and call centers "offshore." Today, in the Fourth Wave, we are trying again to tell people something very simple. More change is coming. That change will be more dramatic and fundamental than the three waves that have come before. Change is indeed exponential, and the rate of acceleration is increasing. To all you mathematicians out there, we are dealing with a fourth order exponential.

Human society has taken millions of years to evolve to its current state of development. About 4.5 millions of years ago, humans were hunter-gatherer nomads, with only rudimentary "human" skills. Less than four million years ago there was "Lucy," who was quite a bit smarter and used more tools. Still, for millions

---

[6] Futron, SIA Report for 2009.

[7] Raymond Kurzweil, *The Singularity Is Near: When Humans Transcend Biology*, (2005) Viking Press, New York.

of years, our ancestors did not evolve very much, and life was about the same. Even 100,000 years ago things were not very different. Today, parents find it hard to understand or penetrate the world of even their own children. People over 40 often find it hard to understand the world of Twitter, hand-held devices with a bewildering array of "Apps" and continual multi-tasking and innovations such as crypto-currencies and block chain security systems.

It actually was only about 10,000 years ago that "humans" began to master the art of farming and to live in permanent towns and cities. This first wave of change from nomadic life to farming communities took thousands of years. Not only did this change the way that people lived and survived, but it also led to a new economic system, first based on bartering and then came the key breakthrough – money. The use of money allowed new skills to develop. Money and towns and cities all served to allow specialization to occur. In time, not everyone was a farmer. There could be bakers and soldiers and builders. By the time of the Chinese, Grecian and Roman civilizations hundreds of professions had evolved, and economic systems had become much more sophisticated. Trade grew between different towns, cities and even nation-states.

It was during the Renaissance that the first vestiges of a true industrial age occurred. The first stages of what we call modern industry and industrial machines began to emerge in the seventeenth century. But, it was not until the nineteenth century that industrialization led to a truly major shift in employment.

In the United States, from around the 1880s to the middle of the twentieth century – over a 70-year period – about seven million people shifted their employment from farming and mining to a myriad of industrial jobs. This same type of shift also occurred in a similar pattern in Europe in much the same time period. In a reasonably short period of time industrialization spread to Asia, South America and Africa. This Second Wave – the Industrial Revolution – occurred much faster and involved many more people around the world, largely because there were many more people. Thus, the nineteenth and twentieth centuries are often considered by historians to represent the Industrial Age, although industrial employment peaked in the 1960s throughout countries such as the United States, Canada and Europe.

From the mid-twentieth century up to the present we have seen the Third Wave as defined by Alvin Toffler in his book by that same name in 1980. This Third Wave was the shift of prime employment in the most economically advanced countries from an industrial base to that of a service economy. This wave of change has now unfolded in at least the United States, Europe, Canada, Japan, Australia, New Zealand and places like Singapore, Taiwan, etc. The nations of the so-called Organization of Economic Cooperation and Development (OECD) now have 70% or more of the work force employed in providing services from transportation to banking, from lodging and restaurant services to accounting, from health care to undertaking, from retailing to amusement and recreational services.

These countries of the OECD have actually seen manufacturing jobs decrease every year since the 1960s. Today many more economically advanced countries have also made the transition to where more than half their nation's employment is in the service sector as opposed to farming, mining or manufacturing. During much of this

Fig. 2.2 The highly automated Kia auto manufacturing plant in West Point, Georgia. (courtesy of Kia Motors America, Inc.)

transition from industrial to service economies, the most advanced countries did well because services are at the high end of what economist call a "value added" ladder.

As the rest of the world catches up by automating its industry and moving toward becoming service economies, the world's riches are likely to shift along with this global trend. Former World Bank president James Wolfensohn describes this almost tidal shift in the world economy thusly:

> *The G-7's dominant role in international affairs over the past half-century was explained by its collective economic weight. Between 1965 and 2002, it accounted for a remarkably constant share of global output – about 65 percent.... Its share [as of 2008] has fallen to 52 percent. By 2030, it is likely to be down to 37 percent, by 2050 to a mere 25 percent.*[8]

In the United States the number of manufacturing jobs has declined with a peak of perhaps 45 million jobs down to below ten million jobs today as America's economy has become dominated by service jobs. Today 85% or more of the jobs within the countries of the OECD involve some form of service as opposed to farming or manufacturing jobs. And this is not just a phenomenon of the OECD countries. A rising percentage of jobs even in industrializing countries – such as Brazil, China, India, Malaysia, Thailand and Indonesia – are now in the service sector. In the seventeenth and eighteenth centuries China and India represented a sizeable percentage of the world economy. Today they are quickly emerging as major economic producers, and as they shift from agriculture, mining and industry to service-based output, their GNPs could rise to be the most significant in the world and even ultimately outstrip the United States and Europe. This shift is why the G-20 has recently come to replace the G-7 as the world's most important economic forum.

With over 85% of jobs in the United States and most of Western Europe now in the service sector the traditional way of looking at jobs becomes less than helpful. In these countries, the "primary sector" of farming and mining typically constitute 3% or less of the workforce. The "secondary sector" of manufacturing claims only about 10–12%. Thus we end up with over 85% in the "tertiary sector" of a catchall category of services. When everyone falls into the same category, the category itself becomes useless. Automation will likely reduce manufacturing jobs within the OECD to under 3% within the next 30 years (see Fig. 2.2).

The rapid drop in agricultural, mining and industrial jobs in developing economies has led to attempts to create new work categories. One attempt has been to define a fourth type of job category beyond the third jobs sector known generically

---

[8] *Washington Post*, Nov. 14, 2008, P. A19.

as the "services sector" – i.e., a "quaternary" job sector. This quaternary job grouping relates to all the positions associated with the entertainment, resort, hotel and
restaurant business. Certainly this sector, that continues to grow within developed
economies, is a quite useful sector to monitor in terms of expanding jobs and its
contribution to the gross national product. Also called the "hospitality industry"
(variously defined to include a wide range of entertainment and hosting industries),
does in fact represent today about 25% of the jobs in the United States and Europe
and close to a third in France.

If one attempts to perceive the "big picture" view of what is happening, the
concept provided to us by Alvin Toffler seems to work well. Toffler, in *Future Shock*
(1970) talked about "High Tech/High Touch." This is to suggest that high tech, and
especially information technology, and computer and artificial intelligence-related
activities, will replace many of the traditional service jobs. Many service-related
jobs in transportation, banking, accounting, etc., may start to phase out and become
largely obsolete. At the same time things that involve "high touch," with lots of
interpersonal contact, may become a much more important source of employment.

These high touch services could be everything from retail sales to restaurant
services, from prostitution (or the now fashionable "sex worker") to nursing, or
from flight attendants to Disney World employees. These jobs will be ones where
machines can play a role, but people will still be needed in terms of human
sociability and our need to relate to each other. Although some such high touch
jobs are today high income – such as doctors, dentists, and lawyers – the majority
of these jobs may evolve to represent lower levels of skills and lower pay. Nursing
aides, teaching aides, retail clerks, bank tellers, driving instructors, workers in
amusement parks and restaurant waiters are high touch jobs, but their compensation
levels are low. Workers in automobile manufacturing who had high-paying jobs on
the assembly lines and who have been retrained to work as nurses or teachers or
other service jobs have actually found that their new jobs paid considerably less.

Even in the service sector we find that automation is increasingly taking its toll.
Bank tellers are being replaced by ATM units. Nursing aides are being replaced by
medical sensors and monitors. Retail clerks are being replaced by scanners that use
UPC strips or magnetic codes to "automate" services for consumers shopping at
grocery or hardware stores or checking in for a trip on an airline. Ironically one
finds that "service centers" are becoming an oxymoron. They are being constantly
redesigned so that personal service can be either reduced or even eliminated. If there
is a megatrend at work today within "service economies" it is to design machines
that can replace increasingly expensive labor. Even jobs such as pharmacists, real-
estate and tax appraisers and graphic designers are being replaced by algorithms,
product codes and RF ID sensors. Only the high touch employees of the Mustang
Ranch in Nevada seem to be exempt from automation – at least for a few more
decades to come.

In the Third Wave, as described by Toffler, a huge number of people have now
shifted their jobs from agriculture and mining (the so-called primary sector) or
industrial jobs (the so-called secondary sector) into service jobs (the so-called third
sector). Tens of millions of people shifted their employment to service jobs in the
last 30 years – and within a much more reduced period of time than during the

transition from farming to industrial jobs. Toffler noted the degree to which the pace of change has increased, and the demands of education and training to adjust to this shift have become ever more rigorous.

The concept of these shifts in employment, occupational training, and transformation of economic systems as summed up by the metaphor of waves coming into a beach, however, tends to mislead us. The timing and the amplitude of these waves are strikingly different. If the first wave took an hour to come to shore and was a centimeter high, then the second wave came in only a few minutes and was perhaps 10 centimeters high. The Third Wave came crashing in within a minute and was several meters high.

So What About the Fourth Wave?

All that we know is that the height of the Fourth Wave is going to be higher and duration will last only a few seconds. The biggest mystery of all is what new jobs will be available and what type of education and training will lead to meaningful employment in this world that arrives on the scene in a very big hurry. Things of all sorts will be in chaos. What things will be stirred up as if in a mixing bowl? The list includes just about everything, but things that will be in limbo include income tax policies, political processes and elections, governmental services and answers about how to pay for social benefits or other governmental services or infrastructure. All of these things and more will have to rely on economic systems that have yet to be invented. Trucks and cars may be driving themselves. Companies such as RedFin will sell property online for a 1% commission that leave traditional real estate sales businesses that expect a 7% commission wondering what hit them. All types of businesses, including instructors and college professors, lawyers, doctors, health care providers, truck drivers, bankers, insurance agents and more will find that their 'secure job' is secure no longer.

For most people the Fourth Wave will seem like a tidal wave – if not a lightning bolt. Jobs will disappear quickly. Structural shifts in the job market due to technological under-employment will not only create stress and turmoil but will challenge us in a fundamental way. It will strongly impact our conventional views of capitalism and complicate efforts to deal with other "mega-crunch" challenges of the twenty-first century, like climate change, to accelerate our concerns about population growth and complicate today's concerns with super-urbanization.

What is driving super-urbanization today is the idea that one moves to the city to find jobs. What do people do if one finds that cities are no longer the place to find jobs, but in fact jobs in cities are the ones that might be reduced the most dramatically and swiftly?

The twenty-first century is going to be not only a time of big time change, but conventional wisdom about jobs, employment, retirement, education, health care, population growth and economic and taxation systems will be challenged as never before. Without some new ideas and new economic and political reforms we are going to be in some serious trouble.

How do we know this? Simply because we are already seeing the first elements of the Fourth Wave crashing onto the shores of America, Canada, Japan, Europe and other countries. And emerging industrial giants such as China and India will find they have an enormous amount at stake as well. The suppliers to these countries will be impacted, and these nations will see the consequences as well.

What is coming next will hit us hard. The next 30–40 years are going to be a huge challenge. We, and particularly politicians, will keep changing the focus as constituents keep demanding action and ready answers where none will exist.

Yes let's save the planet, let's diminish greenhouse gases and let's try to create sustainable economies, but what about my job and a meaningful occupation? If I don't have a job, an income and a meaningful role in society, I suddenly start being selfish, and my sense of social responsibility erodes. Suddenly we could see a breakdown in the social fabric if we do not begin to work out an endgame strategy to take on the big three issues of population control, climate change and employment in the Fourth Wave economy as one vast, interrelated and nearly intractable Gordian knot of problems unique to our times. The challenges are like a perfect storm of problems. We need to act now, but it is still not clear what do to.

People are living longer, but retirement ages have not changed. There is a need for more health care for retirees, but social security programs seemed headed for bankruptcy. People have to be retrained for new jobs due to super-automation, but what are the jobs that make sense for them to be retrained for – especially if they are 55 years old and only some 10 years away from retirement?

Let's spend more money on health care and education and job retraining, says the liberal community, but where are the tax revenues coming from to pay for these expensive programs? Oil spills are hurting the economy. Do something. But we don't have enough gasoline. Round and round the problem will go. Each crisis will distract us from the fact that there are really too many people, and some countries are growing at annual rates from 2% to 6% per annum. One truth is apparent. More people and burgeoning urban populations are clearly a significant part of the problem. If we could only return to the world population of the 1950s of around four billion we could solve many of the issues much more easily. The fact that we will have a global population of between 10 billion and 12 billion, and perhaps 50 megacities of over ten million residents, by 2100 is a key problem of our times. The need for more housing, more food, more education and health care and more jobs represent huge problems. All these new people will generate a good deal more pollution, accelerate the climate change dilemma and also exacerbate the employment problem in the age of super-automation. These three interlinked hyperobject type problems seem well beyond the grasp of today's world political and economic leaders' capabilities.

The sad truth is that we will have no viable economic, political and regulatory programs to do all the things we need to do all at once to respond to these various crises. There are no truly effective ways to get nations and world leaders to begin needed reforms and actions at this time.

Let's put things in perspective. The Fourth Wave is really going to be like a tsunami. And it is going to hit us fast.

Why are we going to be unprepared for this change? First of all, people tend to ignore what could be potentially bad news or disruptive forces for as long as possible. Only if one has access to the global media headlines are such warnings posted, and the global media has a vested interest in not explaining the mega-crunch forces that are coming. Alerting people that problems are coming is bad for business.

These new forces of change are everywhere about us. They include expert systems, artificial intelligence, water shortages, greenhouse gases and polluting energy systems that are the by-product of a thriving global economy and a burgeoning global population. These disparate forces are all instruments of Fourth Wave changes. These are all issues that we can easily choose to ignore or at least pay little heed to because they can simply be forestalled, as next generation's problems.

What is the single largest source of pollution and greenhouse gases on the planet today? Is it coal-fired electricity plants? Is it SUVs, or gas-guzzling clunkers or jet airplanes? Is it oil spills or a lack of recycling? No. Although these are clearly major sources of pollution, the number one source of climate change is simply human population growth. Each new child born within the United States will generate thousands of tons of greenhouse gases. It is possible that even with birth control and greenhouse concerns the human population will grow to some 12 billion by the end of the twenty-first century.

This projection is the single most threatening trend to human survival – outstripping nuclear war, biological-chemical warfare, racial genocide and terrorism. But this continued human population explosion has other potentially devastating implications beyond climate change. Super-automation (the coming Singularity) will create structural unemployment and technological underemployment with a vengeance. In short, in an age with artificially intelligent and self-aware machines, like HAL from the sci-fi epic *2001: The Space Odyssey*, we will need fewer people, not more. More and more people on a resource-limited 6 sextillion-ton mudball we call Earth is a luxury we cannot afford. Where does the human population peak, and perhaps begin to descend? Is it 10 billion? 12 billion? 15 billion? 20 billion? Can we survive with that many people? Where will the energy come from? Where will the jobs be to support that mass of humanity? In short, the key to the coming Fourth Wave will be to become more educated, live more sustainable lives, have fewer kids, and create a less rapacious economic system to let people and our planet live in a more healthy way.

Most people are content to say population control will take care of itself. This convenient assumption is that the constraint will be food and water. In fact, it will be climate change and the loss of jobs due to super-automation that are most likely to lower the boom. Water and food supply may indeed become a problem, but a curtailment of population could be one of the most efficient ways to address a food and water supply problem.

Overpopulation, climate change, super-automation and radical changes within our Internet-dependent world and new ways of social networking on a supranational scale will come together over the next two decades in ways that will force on us fundamental change. These drivers of global change will shake us up in ways we are ill prepared to respond to. At risk will be national security, food, water and energy sources, climate catastrophe, technological unemployment on a massive scale and global economic instability.

So what are our options? Diligent monitoring of both positive and negative trends, pro-active planning, economic and social rewards for moderation of extremist views and actions, new types of economic incentives to reward responsible behavior coupled with disincentives and penalties for extremism must all become a part of our lives. The place to start is to have a better sense of where we are going.

Monitoring of the so-called DEGEST factors are perhaps key. We need to be able to model and understand the interaction of: **D**emographics, **E**conomics, **G**overnment, **E**nergy and Environment, **S**ocial, cultural and religious factors and **T**echnology. These interactive forces of change create a rich brew. Survival of a global society now depends on how each of these six factors impact the world. Complexity continues to increase as a function of human society and current trends[9]:

(a) **Demographics.** More people make everything more difficult.
(b) **Economics.** Higher economic throughput and more trade and higher volumes of transport create not only more complexity but more pollution, more urbanization, more interpersonal conflicts and greater security threats. As long as growth and wealth remain the prime drivers of economics, as opposed to survival values, the dilemma will grow.
(c) **Government.** More governments and political systems, each with complex laws, all interacting with one another, will compound problems of dealing with global pollution, security problems, unemployment, fair trade and remedial actions.
(d) **Environment and Energy.** More environmental damage, more energy consumption, more greenhouse gas release, more species loss and increasingly rapid climate change will create a deadly equation that can only be changed by a combination of technology, regulatory change and revised economic systems.
(e) **Social, Cultural and Religious Belief Systems.** These increasingly complicate all forms of human interaction. Extremism, social zealotry, terrorism and religious fundamentalism create a dangerous brew to modernist societies.
(f) **Technology.** Everyone hopes that technology will make the human condition better, but, in fact, technology makes life more complex, with increased societal vulnerabilities that accompany each and every innovation. Technology is driven today not by survival values but economic growth values. Until we find a way to get the motive force of technology behind survival values we will remain in peril.

The bottom line is that global society and human civilization are in peril. The perils are unfortunately almost everywhere. Perils in the twenty-first century are not hard to find and actually are easy to identify. The following perils and related dangers are only the top ones we seek to address in the pages that follows.

---

[9]Futuring.

Twenty-First Century Perils that Need to Be Closely Monitored

- **Climate change.** This is one of our greatest threats.
- **Destruction of the ozone layer and deadly cosmic radiation.** Genetic mutation from solar and cosmic radiation could quickly end the human race – much faster than climate change.
- **Pollutants.** Various pollutants include poisons, radiation hazards, various hazards to health, change to Arctic cap reflectivity. All can lead to species extinction.
- **Super-automation.** This can be a source of technological unemployment, the need for accelerated job retraining and educational needs, national conflicts due to trade disputes, fewer jobs, etc.
- **Continuing patterns of population growth and lifetime extension.** The increase in population, the increase in lifetime expectancy, along with extended periods of retirement are the big threats and are key factors in climate change, pollution, unemployment, and runaway education and health care costs.
- **Super-urbanization and rapid globalization.**
- **Social networking and unrestrained global interconnectivity** (i.e., the Internet, satellite and cable television, talk radio). Can lead to unlimited and unmediated streams of child pornography, racist and hate literature, and political extremism.
- **Accelerating education, health care and retirement costs**. These increasing costs plus the unknown consequences of automated health, education and training systems need to be closely monitored.
- **Terrorism, extremism and rising costs of urban and global security.**
- **Complexity**. There are many unanticipated social, economic and political consequences that can flow from an advanced technological society. We must find better ways to cope with these factors and their potentially very dangerous impact on our future.

Unfortunately, many of us will not like the information and social implications that can flow from the new knowledge derived from the monitoring of global mega-trends. We are truly unaware of many of the economic, social, cultural and political implications that now flow or will flow in the future from the use of new technologies. Clearly there are downsides from a surging global population, but to marketers this continues to be seen as a bonanza of market expansion. More people means larger markets for businesses that want to sell more products and services. But more people means more dramatic climate change and the potential loss over time of all people on the planet and thus the eventual end of all markets. Two overly simplistic equations spell out the difference between short-term and long-term futures. In the short-term more people can increase markets and profitability. In the long-term more people may well serve to end all intelligent life as our planet heats up out of control and is saturated by a blanket of dense gases that destroys all life, like on Venus. Today Venus is a planet whose atmosphere burns up metals and is so toxic that no life is possible. Before runaway heat enveloped Venus in a shroud of greenhouse gases, this near planet might well have been a paradise that sustained life, but today it remains a deadly symbol of what might happen to Earth.

How do we cope with a "complex" future? We have a much better chance of coping with climate change, super-automation, super-urbanization and societal complexity if we could do just one thing. This would be to reduce human population. If we could only turn back the clock some 50 years to recapture the world of our grandfathers and grandmothers.

If we could somehow suppress the ticking population bomb of our times we would have the opportunity for a new lease on life. Reduce our population two to three billion, and we would have a much better chance to cope with all these other issues.

A global population of say four billion would likely produce half the pollution, eat half the food, allow our cities to be much less crowded, unemployment could also greatly shrink and many other benefits would flow. We would actually have all the genetic diversity we need in a gene pool of only 100 million or so people. A free-enterprise, capitalistic society is great at producing more and more economic throughput. But the system that brought us more technology and higher living standards turns out to be lousy at coping with today's climate catastrophes or adjusting to super-automation, rapid-urbanization or a complex future. We have a great challenge before us. That challenge is to contain the economic, social, cultural, political or demographic complexity of a twenty-first century world and somehow change it to our advantage. The twenty-first century is crunch time in dealing with the unique challenges of our time.

What are the answers for the future? Is it more birth control? Is it greater rewards for reduced family sizes or a tax on families with over two children? Can we survive by relying on more sustainable and recyclable environmental and energy technologies? Would higher environmental taxes allow us to make a more rapid transition toward living more sustainable and "greener" lives? Can we invent future cities that allow us to create our food within the urban environment? Can we invent clean electric cars? Can we design improved transport, IT and telecom networks that make our service economies not only more efficient but ecologically sound as well? Can we invent a new form of a restructured free enterprise economic system that maximizes the value of sustainable living and human survival over raw economic throughput? Can we find ways to turn social, cultural and religious beliefs to our advantage, especially when it comes to survival of the human race?

Perhaps the answer will ultimately be almost all of the above plus new ideas and new technologies, systems, regulatory policies or economic incentives yet to come. We may need to rely on what the best and brightest of today's millennials will be able to think of and create in order to fight through the Cyber Revolution that will come in full force in just a few years' time.

Some believe that the hope for the future can come through new and better forms of technology. Artificial intelligence guru Ray Kurzweil has suggested that the Singularity will produce a sort of panacea. Kurzweil has suggested that super automation will bring a sea change in every conceivable way. The human database of global knowledge that is now doubling every year to 18 months will zoom forward and perhaps expand a 1000-fold in a single year. This mushrooming human intelligence could then be employed to create clean energy, sustainable lifestyles and much, much more.

Others envision that new types of human systems and "smart corporations" such as Google, Oracle, NEC and IBM, may be able to succeed in "reinventing" our world in ways that outmoded governmental political processes have never been able to achieve. The point is that invention, technology and innovation will need to pursue the interests of human ecological destiny rather than those of narrow corporate profits. Breakthrough science and technology without new goals and objectives will not succeed in meeting twenty-first century challenges.

The great twentieth century inventor, philosopher and innovator R. Buckminster Fuller had a comprehensive view of the human condition and even included a special role for humanity in the universe. He was among the first to suggest that human innovation was needed to balance the force of entropy in the universe. In essence, Fuller posited that the human mind – at least within those parts of the universe we now know – was uniquely positioned to achieve what he called "ephemeralization." Fuller boldly suggested that the human mind was the anti-entropic force in the cosmos. Entropy, of course, is the tendency of all systems to wind down and become disordered. Fuller observed that there is hot and cold. We also have positive and negative, i.e., electrons and protons. We even have matter and anti-matter. Thus the "one-way nature" of entropy bothered him a great deal. He suggested that "intelligence," which can create order out of disorder, must not be a freak accident but something essential to the universe. R. Buckminster Fuller – never a timid soul – pulled a rabbit out of a hat and pronounced that there is human science and technology because the universe would ultimately wind down to total disorder and chaos without it. *Cogito, ergo sum.* Thus we not only exist, but we are a force of nature in this great universe of ours.[10]

The human mind, at least in our neck of the cosmos, is uniquely able to create nuclear fusion and build complex systems. Perhaps in time we might ultimately be able to re-engineer a planet or explore the vastness of space in interstellar spaceships. Certainly, it is now within our intellectual purview to conceive of how to build solar shields to protect Earth, and even Mars, from coronal mass ejections (CMEs) and solar storms.

It is a great power that humans now have from their fast-developing technology – for both positive or negative change. The first and foremost challenge of the twenty-first century will be to create new systems and practices that can allow a livable biosystem to sustain itself on Earth. The same force, if we can survive as a human race, could allow us to colonize the Solar System or even the galaxy, but first we need to find a way to survive and not transform Earth into a dead world. We are, in short, currently on the narrow cusp of success or failure.

In essence, we could find the willpower and the technology to create a sustainable form of life on Earth. Or we could manage to fuel runaway climate change,

---

[10] R. Buckminster Fuller.

where something like 12 billion people find out in dramatic fashion what limits to growth really means. We could indeed manage to accomplish through climate change what a barrage of H-bombs might not even achieve – the total disappearance of future generations of humanity –possibly as soon as the early twenty-second century. *Homo sapiens* could end as a failed experiment.

Perhaps, as humanity fails, some other form of intelligence of superior quality may succeed elsewhere in the universe. Even if Buckminster Fuller was right about the need for anti-entropy to exist, there is nothing to say that human intelligence has to be the only successful form of thinking and organizing matter. This challenge is unique to twenty-first century Earth and to twenty-first century humanity. We can thus succeed or fail, based on decisions we take individually and collectively. Ultimately it comes down to whether humanity can cope with the complexity that comes at a time that Arthur C. Clarke once eloquently described as *Childhood's End.*[11]

We need to invent better cities, create new economic incentives to control population growth and give higher value to a sustainable global environment. We also need better education and health care systems as well as better and greener energy and environmental systems that allow not only humans but all types of flora and fauna on our 6-sextillion-ton planet, to survive and even thrive. Finally, we need better strategies for coping with a networked world and innovative new ways to approach world peace in an international environment that is constantly inventing new weapons of mass destruction and gives rise to terrorism and religious and political extremism.

Is this a very tall order? Yes, indeed it is. But as Albert Einstein once said: "The questions are the hard part." Until we put together a creditable list of what ails us, we cannot put together a creditable action agenda to move us in the direction that we not only need to go but must go if we are to survive as a species. As Charles Darwin wisely noted after decades of study, the species that survive are not the smartest or the strongest but those that are most adaptable. Today the most depressing observation that we can make about *Homo sapiens* is that despite our clever technology and amazing developments, our economic, political and social systems are incredibly rigid. In our exercise in "futuring" and anticipatory planning, we must find a way to adapt. In an increasingly commercial and capitalist world that seems to imply that we must find a way to reward positive change and economically punish unnecessary non-adaptive behavior. This means, for instance, rewards for recycling and fines for not doing so. It means, as another example, a loss of tax benefits for having more than one child and tax increases for having more than two.

The motive behind explaining these twenty-first century super trends that will bring the Fourth Wave crashing down upon us is not so much to alarm the people of the United States and the rest of the world, although some alarm and respect for the power of what is to come is indeed desirable and healthy.

---

[11] Arthur C. Clarke, *Childhood's End.*

No, the real purpose is to bring new meaning, understanding and even some acceptance of the great "complexity" that a Fourth Wave society brings to our world. Today, this is a world with too many people, with economic and social systems that are ill-equipped to deal with exploding technology, and a populace that is simply not ready for what is yet to come. In the words from Arthur C. Clarke's book *Childhood's End,* the question is, what do we humans want to be when we grow up?

We could become responsible citizens with a destiny that extends to the Solar System, the galaxy and beyond. Or we could become yet another failed experiment in biological evolution that imperils not only humanity but a great many other species as well. There have been at least six times in known history where mass extinctions have occurred. This means a time in which more than one third of all species succumbed to some mega event over the last three billion years. It is possible that the first "human-triggered" event of mass extinction could occur within the next hundred years. The choice for the very first time may well be up to us.

Chapter 3
The Prime Drivers of Change

A hyper-object is an event like climate change. It is so massively distributed in time and space relative to humans that it is hard for us to grasp.

Tim Morton, environmentalist

Exploring the Prime Drivers of Change

As already discussed in Chaps. 1 and 2, the clock will run out soon on a post-industrial global human society as we know it today. By 2100 the world will be greatly overpopulated, and as a result water, food, jobs, and social and economic stratification will all constitute serious problems.

By 2100 climate change trends may perhaps have reached a point of no return or at least become a matter of dire concern. In some parts of the world this is part of today's reality. In Cape Town, South Africa, you can see people dancing to celebrate a rain storm that is finally providing some relief in this drought-stricken city. In this lovely city we are being told not to take showers, and to flush toilets only when we must. The University of Cape Town has set the goal of reducing its water consumption by 50%.

By 2100 we may very well have a global employment crisis. Further, this will also be accompanied by an environmental and sustainability crisis. This would be triggered by the dual threat of continuing overpopulation and super-automation. A recent study by the McKinsey Global Institute has found that in 60% of the industries it studied at least 30% of all jobs in those industries could be replaced by AI today – not tomorrow but right now. Of course as we approach the future, AI will become more and more capable and more and more jobs will be a risk.

© Springer Nature Switzerland AG 2019
J. N. Pelton, *Preparing for the Next Cyber Revolution*,
https://doi.org/10.1007/978-3-030-02137-5_3

The economic and employment implications of super-automation cannot be over-stated.[1] Nor is this the only negative aspect of super-automation. There are also issues related to the loss of privacy and the 'weaponization' of information derived from social media that can threaten democratic values and be used to manipulate elections.

All three of these interrelated twenty-first century challenges may reach a crisis point unless we make some basic changes. Prince Charles of the United Kingdom has suggested we only have 100 months – less than a decade – to reform how we live, work, consume, and procreate before Earth's biosystem starts to spiral out of control, as Venus did many years ago.

The good news is that the window for environmental reform is a bit wider than the next hundred months. The bad news is that the longer we wait to ameliorate climate change challenges the higher the cost. The window to environmental reform may be open for perhaps another 50–80 years. Yet, the longer we wait to create truly sustainable energy systems and to curtail the growth of human population, the higher the cost. The bottom line is that there is a limited time to act to save humans – and many other life forms – from mass extinction and economic crisis. This sometimes seemingly inevitable crisis for humanity even has a name. We now live in the Anthropocene age, and we are apparently headed toward the sixth mass extinction event that has been given the name the Anthropocene extinction. It is clever of us to have identified the causes of the next mass extinction – overpopulation, a disposable economy, and climate change – yet we seem unable to act to preserve our species because of global economic thrust. The key question of our time is can we create new economic drivers and growth vectors that can save ourselves from the Anthropocene extinction?

That window of survival is not flexible. Our opportunity for longer-term survival is not open-ended. Dinosaurs lasted for hundreds of millions of years. Humanoids may bite the evolutionary dust in just a little over 4.5 million years. And dinosaurs were supposed to be a lot dumber than we are. Time will tell which species gets the award for survival. Humanity today does not have very favorable odds by any knowledgeable odds-maker. Any moxie odds-maker in Las Vegas would likely give the current odds of the Anthropocene extinction occurring as being 10 to 1 over that of long-term human survival. There can be little doubt that current administrations in the United States have moved the odds of survival strongly in the wrong direction.

What can we do to prevent a mass extinction event? How can we marshal the forces of economic growth to drive change in a positive and 'green' direction and to change a disposal economy into a circular economy, where there is a zero carbon footprint for our cities, vital resources are recycled, and economic growth is steered

---

[1] Lisa Morgan, "9 Ways AI and Intelligent Automation Affect the C-Suite" Information Week, July 27, 2018, https://www.informationweek.com/strategic-cio/9-ways-ai-and-intelligent-automation-affect-the-c-suite/d/d-id/1332383?_mc=NL_IWK_EDT_IWK_review_20180727&cid=NL_IWK_EDT_IWK_review.20180727&elq_mid=86000&elq_cid=20198565

toward new enterprises, products and services that drive us toward survival rather than extinction of life on spaceship Earth. The biggest question of all is how can we reign in population growth? A global population that exceeds 10 billion is truly a ticking time bomb. All the green policies in the world can't cancel out billions of new people on our planet.

The Exciting New 'Smart Industries' That Can Help Save Our Planet

Deceased astrophysicist Stephen Hawking and space launcher wizard and electric car enthusiast Elon Musk have said words to this effect: "There is no Plan B, because there is no planet B."

However, there is room for some hope that we humans will get our act together and pull off a better Plan A to find a way forward to create a sustainable habitat right here on spaceship Earth. If we are clever enough to start necessary reform, we might even design new environmental and space-based sciences and industries. Yes, these new industries will make a profit and do the things that companies do. But we can hope they will also create new and better tools to fight the big three hyper-object threats outlined and discussed in Chaps. 1 and 2. For better or worse, we are likely to depend on the NewSpace and cybernetics companies, which some have called Space 2.0 and Cybernetics 2.0 companies, to save Earth and human life on our planet. These new enterprises are going to create new opportunities both here on Earth and off-Earth as well. If we are lucky some of their innovations may be able to help save life on our planet as we know it today.

These space-based enterprises. as well as cyberspace businesses, can help us re-invent our patterns of living and create more sustainable practices in business, governments and neighborhoods across countries and around the world. Such new initiatives can provide new employment, more environmentally friendly practices and generate wealth. These new industries, if well conceived, could even help save our planet from runaway ecological destruction and perhaps economic, business and under-employment trauma as well. Further, part of our future might depend on creating new structures or systems in space to let our new, greener systems fully kick in and start zero carbon-footprint cities. Our survival may, in part, actually depend on building new types of mega-structures in space to protect Earth against cosmic hazards.

One such large-scale space survival tool would be what has been called LAPSE. LAPSE stands for LAgrange Protector from Solar Ejections. At one end of the scale Space 2.0 technology would create protective shielding for modern global infrastructure against cosmic hazards and at the other end would create new enterprises to generate new sources of clean energy, monitor pollution, or bring Internet connectivity and health and educational systems to underserved regions of the world.

Cybernetic industries 2.0 will have an even more critical role. These new green will need to lead the charge forward in developing and implementing the critical sustainability technologies. This means the development and use of smarter and greener electronic technologies that will be used within new enterprises. They will be dual-use enterprises that are key to saving the world from mass extinction and also help provide new jobs and in the process also meet many of the goals of the U. N. for Sustainable Development (See Appendix 1 at the end of this book).

There is an old saying that goes "when life gives you lemons, make lemonade." Clearly there is a great deal of wisdom behind this adage. Oddly enough overpopulation, the need for new jobs, and climate change all demand a Fourth Wave, or circular, economy.

Recently ex-governor of Virginia, Terry McAuliffe, noted that by the time he had left office he had helped to create 50,000 new green energy-based jobs and that the number of coal-based jobs has shrunk to just over 1000. This is the type of results that are needed around the world.[2]

Today's dire need to save the planet also represents a significant business opportunity. The wave of the future is to create new tools that can make the world cleaner, safer and work less arduous and less dangerous. This would also result in a new generation of people that are better educated and healthier, and are able to live longer and more fulfilling lives.

What would be the nature of these new Fourth Wave companies? Some would make better and lower cost solar cell arrays or quantum dot sustainable energy systems. These industries would serve to replace coal mining. In this smarter and better world, coal miners and those that are currently working in such dirty, polluting and dangerous professions would be recruited and trained to undertake these new jobs that would safer, cleaner, less dangerous and hopefully better-paying.

There could be workers recruited to help manufacture electric cars and even electronic aircraft to replace cars and planes that run on petroleum products. If we embrace, as a part of a longer-term plan, the problems of combating climate change, overpopulation, and possible future under-employment chaos, the potential for prosperity and greater affluence can greatly expand within a single generation. There is the potential of creating important new industries with new jobs and products and services that we need. The technical and scientific capabilities to create clean and sustainable cars, planes, transportation systems, offices, energy systems and even agricultural, food and water supply systems are largely known. Such new industries, products and services could fuel a world economic boom. What is lacking are political, economic and business leaders who have the foresight, intestinal fortitude and strength of character to transform the world sufficiently that we don't end up destroying Earth.

The way forward does involve some unprecedented steps that are hard. Governments need to buy up, shut down and retool the world's biggest polluters – both polluters of the atmosphere and the oceans and waterways. This means buying up all of the coal-fired plants and taxing or creating financial incentives to convert

---

[2] Meeting of Georgetown University alumni at Gov. McAuliffe's home on June 4, 2018.

cars, trucks, buses, tractors, various other types of machinery, boats and other petroleum-fueled vehicles to electric or other clean and sustainable technologies. And this means a universal program that is carried out not in just one or two countries but in virtually all countries on an expedited basis. Part of the plan would be to recruit the workers that are engaged in mining, coal power plant operations and other such jobs and retrain or redeploy them in new and better jobs. If this involves the resettlement of workers, there should be incentives and tax benefits for those directly involved in terms of education and health care services, and even support for relocation costs.

We need to clean up all the various types of machinery that we use and even deploy new space systems in new ways with new, innovative designs. This might involve the building, deployment and operation of a number of newly designed solar power satellites. Such new systems would be able to deliver clean energy where needed to various sites around the world. It might involve creating more wind farms, ocean thermal energy conversion plants and the creation of new transportation systems that are cleaner on one hand, or even more logically, there should be a conversion of more jobs to telecommuting-based activities. It is always cheaper, cleaner and more efficient to move ideas and thoughts than people in and out of cities. In places like Madrid, where there are siestas, the movement of people in and out of cities four times a day is a very peculiar idea in a world where most jobs can actually be performed as a tele-service of one sort or another.

The overall equation is not that complicated. The idea is to create and implement more business activities that hire more people to create clean energy, transport, housing and office buildings. In this way old-fashioned and polluting industries are retired on a wholesale basis. This must be designed in a way that can create new jobs and opportunities. Efforts to modernize, increase efficiencies and convert polluting systems should be coupled with projects to create new job opportunities. If this were to be done systematically and cleverly with incentives and where people in the first wave are actually competing to be the first volunteers for these programs, then people would feel recruited to enter a better, cleaner, safer world where health care and education would be seen as a right.

If done well with support from developed economies, many countries from the developing world might be able to parachute – or at least quickly transition – into a newer and better world. This might also include plans for desalination plants, new tele-education and tele-health-based schools, hospitals and clinics and more. Critics will argue that such a transitional plan will be expensive and perhaps create significant debt.

But war, overpopulation, polluted air and oceans, coupled with climate change that gives rise to horrible destructive storms, and larger and larger amounts of greenhouse gases trapped in the upper atmosphere will in the longer run be far more expensive and destructive.

Other goals would include: (i) the redesign of our patterns of consumption to create a sustainable world through recycling and rejection of a disposable economy; (ii) control and limits on runaway urban population growth, either through tax incentives, rewards to towns and villages or other types of carrots and sticks. There are many other aspects of a transition plan to make the world more sustainable, greener, safer and less prone to war. What is clear is that the number of humans inhabiting spaceship Earth is rising too high. The amount of air and water pollution is too great. The level of throughput of products and services is creating more problems than progress. These goals might also be reformed to include new economic programs to eliminate excessive consumption, excessive resource depletion, and more efficient pricing to encourage sustainability and discourage the most egregious practices that lead to the worst levels of pollution and thoughtless levels of consumption. Ultimately such plans will also serve to redefine the purpose and patterns of work in this technology-driven world.

The bad news is that this 50–80-year time horizon, our "window of survival," is premised on the idea that all of the industrialized and economically developed world will increasingly implement sustainable development programs and that the lesser developed states will, in time, follow suit. This is probably optimistic at best.

Our ability to survive as a species actually depends on sustainability programs picking up speed from the 2020s. The key is whether most cities will have achieved a zero-carbon footprint by 2050. We hope most nations have the right mixture of carrots and sticks, sufficient to achieve zero population growth in both developed and developing countries by mid-century as well as perhaps an ultimate decline back from the dangerous levels we are now headed towards. If we don't get greenhouse gases under control soon and stave off a swelling global population then all bets are off. The longer we put off achieving these goals, the more likely our planet will go the way of Venus. Incidents like world wars will be small potatoes indeed.

Mega-Deaths or Giga-Deaths?

Controlling pollution and population are just the most important changes. We believe that we have at least five decades to adapt to a new environment in which smart machines and the Internet of Everything dominate the world's economy. In this new world, labor will take on new meaning. We believe that it is possible to learn how to embrace green practices without excessive economic difficulties or chaotic social sacrifices. But only if we begin significant reform NOW.

Many have suggested new paths forward. Microsoft founder Bill Gates has suggested a tax on robots. Others such as political leaders in France have tinkered with retirement ages and a shortened work week. Keynes told us that workers can even be paid to do the equivalent of digging holes and filling them up and a modern service economy can sustain itself. The answers are far from clear, but the whole idea of employment, taxation, economic growth and sustainable development will, in the end, all need to be redefined in the world of super-automation.

The key is simply adapting to a new economic value system. In this new system, survival will have a higher value over growth. Sustainability will become a more critical goal than personal wealth. Billionaires like Bill and Melinda Gates, Warren Buffet and other enlightened business magnates, have urged policy leaders to invest in creating a sustainable world. This cannot be accomplished overnight. The process that moved Copenhagen, Denmark, from a heavy carbon-based fuel consumer to a zero carbon footprint required over 40 years. We need to find a new pathway forward within the time span of the twenty-first century to achieve a livable planet for humanity and flora and fauna across the seven continents and all oceans.

The triggering events for the Fourth Wave are several-fold. The drivers of the Fourth Wave have many dimensions, and this is confusing to economic and political leaders in terms of where to start identifying the most critical targets. One profound source of change is obviously the pending climate change, a deteriorating ozone layer, thawing peat fields in Siberia and oil-blackened icecaps. These disturbing ecological events plus a torrent of industrial greenhouse gases can lead to a climate catastrophe. Action on climate change requires major reforms across the globe within the next few decades. Ecological targets are clearly of prime importance.

This will require the carrot and the stick approach. Clean energy can be incentivized and polluting industrial actions can become felonies or trigger major fines or tax penalties.

The increasing onset of climate change will hit humanity in a host of way. We will see a rise in the sea level, experience increasingly violent storms and suffer through widespread droughts that will impact food supplies and livestock. The biggest problem is likely to be growing lack of potable water worldwide. Clearly the steady increase of people, cattle and livestock are prime contributors to greenhouse gas build up. World population stood at 800 million people in 1800. By 1900 there were 1.8 billion people. Now we are over 7.5 billion people, and dangerous growth continues. The one simple means of rescuing humanity from extinction is simply to reduce the number of people on our planet. Our spaceship Earth can only hold so many passengers. The maximally optimal number of humans is about four to five billion max. Clearly if this were an elevator or an airplane we would have to jettison our overloaded number of passengers. The answer for controlling excessive population growth will likewise involve both incentives and penalties. The goal ultimately is not only zero population growth, but to cut human population in half, if growth patterns actually peak at 10–12 billion. This is not something that politicians or legislators are willing to say. Business leaders with business plans continuously based on ever greater growth are equally unwilling to accept that continuous growth is a death sentence for humanity unless we can develop technology to go safely and permanently into space and colonize other worlds. In the meantime, four billion people is a reasonable and sustainable number for the human population. The best path to survival and sustainability is to gradually reduce global human populations back to the levels of the 1960s.

But climate change and population growth are not the only dynamic drivers of the Fourth Wave world. We have a number of other drivers to monitor and need to understand how these drivers interact – socially, technically, politically and economically.

A further major source of change is what futurist and IT technologist Ray Kurzweil calls the coming Singularity but which we will call more modestly super-automation. Regardless of whatever one calls it, super-automation means the coming availability of self-aware, "thinking" machines that can in various mechanical forms assume the great majority of agricultural, mining and service jobs across the globe. The First Wave changed human society in a fundamental way 10,000 years ago. The transition from a nomadic life based on hunting to a new life based on permanent settlements and farming was the most dramatic change in the four million years of human life on Earth up to that time. We think of farming as ancient, but the age of human agriculture represents only 0.25% of the time of existence of intelligent anthropods, or humanoids.

Agriculture – the most important of human technological inventions up to that point in human existence – led to many other innovations. Farming was the key technological innovation that led to the creation of nation-states, territorial wars and new types of weaponry, the building of cities and fortifications and specialization.

The Second Wave was the industrial age, and it was born of specialization, technical invention and ever greater complexity in human civilization. Over a period of centuries industrial production replaced farming and mining as the prime activity of humans on our small planet. Before the Second Wave was completed across the world, a Third Wave, a new type of economy based on services, replaced industry as the top source of employment in the most economically advanced economies. Indeed, in the countries of the OECD industrial employment peaked in the 1950s, and new jobs such as those in services such as transportation, communications, education, health care, tourism, guest lodging, insurance, government and hundreds of other service jobs became the prime source of employment in the most advanced economies.

The Fourth Wave will be shaped in a primary fashion by basic factors that include climate change, population growth, super-urbanization and super-automation. These four drivers will in turn have profound and dramatic impacts on the world economy. They will, in turn, reshape the direction of technological development, revamp economic and regulatory systems, refine work, reshape social interactions, health care, education and training needs, as well as complicate and restructure the need for security at the local, national and global level.

The Fourth Wave represents a time of turmoil. It requires a charting of trends and monitoring the pathways forward to control pollution and climate change, gradually reduce the global population of humans, adjust to super-automation and the Internet of Everything, and find better ways to cope with super-urbanization.

Coping with these powerful drivers of change will require political, economic, social, and technological adjustments and change as complex and difficult as at any time in human history. On top of everything else, the Fourth Wave will be a period of "future compression," where change will occur at an accelerating pace, and political, economic, social and technical responses will often seem way too slow. How can we possibly put the brakes on these drivers of change?

The question is what happens to all of humanity in the twenty-second century if the end result should be runaway population growth, runaway climate change and massive issues of productive human employment?

This is not a question to pose to political leaders, business magnates, or to under-educated and prejudiced people that might be described as the "Archie Bunkers" of the world. It is a question to pose to scientists, engineers, urban planners, ecologists and dispassionate experts who can help to figure out creative solutions. We will need our very best minds to figure out a viable plan of action as to what to do when we run out of drinking water, food, and key resources, or when Earth's climate becomes too hot, too arid and too hostile to sustain mammalian life forms.

Demographics Is One of the Key Drivers

One of the "brakes" that could be applied to one of the key drivers that would have a broad and positive impact on all the other drivers would be demographics. If we could slow population growth or even better reduce human population growth in the next five decades by a billion or so it could slow everything else. Urbanization would slow. Energy demand would decrease. Unemployment could decline. Health care, education and training needs would be less. Even security and military costs could decline.

The objective of this chapter is thus not only to identify the various key drivers of change but also to explore how to mitigate the most adverse aspects of these twenty-first century change agents – especially those with a harmful impact on human survival. Population growth seems to be enemy number one, and dirty energy may be enemy number two. Enemy number three may be economic systems that do not properly value human survival.

If the world environment is to remain safe for life on Earth, then human population growth needs to be reduced as much as possible while we are also making dramatic new investment in clean energy, sustainable heating and cooling systems and "smart" transportation and IT networks. New "clean and green technologies" need to be developed across the board, and health care and educational systems greatly improved. These are the growth industries of the next five decades.

Super-automation of most jobs will continue to accelerate. Smart machines will increasingly be employed to take over an amazing array of service jobs. These self-aware machines (SAMs) can make super-automation not only possible but also cost effective. Such machines, by not only being smart but also being ecologically clean, can help move us toward a more sustainable economy and can also eliminate the need for quite so many billions of people on the planet.

These SAMs, using a combination of expert systems, artificial intelligence and heuristic algorithms, can not only assume jobs in the primary sector (i.e., agriculture and mining) and the secondary sector (i.e., manufacturing), as they already have, but move on to the tertiary sector (i.e., all types of services). Machines have already become, in economically advanced economies, the prime work force in farming, mining, assembly line manufacturing and textiles. The next step is for smart machines and artificially intelligent software to become accountants, pharmacists, property assessors, teachers, transportation and civil engineers, architects, cashiers, retail clerks, insurance salespeople, actuarial statisticians, public relations representatives, inventory clerks and most of the jobs in governmental bureaucracy.

Today, the Chinese and German economies seem the strongest in terms of employment and growth because they have already largely adapted to industrial production for this new type of economy. The U. S. "market driven" economy is not efficiently adapting to this new world because it has no strategic mechanisms in place that can anticipate change and adapt to it decades ahead of this emerging new world reality.

It can be argued that no one can truly predict the future, and thus strategic planning is often a wasted effort, fraught with peril and with no reliable results. In our current world of future compression – where over 4 exabytes of new information is being produced each and every year – this notion is pure folly. Futurist Edward Cornish, founding president of the World Future Society, argues quite convincingly in his book *Futuring* that if one simply monitors the global trend lines shaped by six drivers of change, future conditions can be largely revealed. Cornish's six drivers are summed up by the acronym DEGEST – Demographics, Economics and business, Government and politics, Environment and energy, Societal and cultural trends, and Technology.[3]

One merely needs to monitor current global trend lines of demographics and human population growth to realize we are in trouble. The population growth alone will drive unfavorable climate change, resource depletion and growing environmental devastation. Even with improvements in green technology and better sustainability practices, this massive growth of population alone will contribute greatly to growing greenhouse gas emissions, massive oil spills, ozone layer depletion and more in terms of environmental damage. We will need to rely heavily on burgeoning new environmental technology to help stem the tide. Only with the population peaking, and perhaps seeing some decline, is there any hope of coping with climate change and unemployment. The stemming of population growth is the single most important "green practice" that could be implemented over the next 50–80 years. The rapid onset of super-automation and unemployment will likewise be exacerbated by this massive population increase. The projected increase in human population by perhaps 4.5 billion people by 2100 (i.e., more than all humans since 4.5 million years ago through 1800) is a most daunting projection. It almost single-handedly documents that the global economic and political systems of the modern world are unstable. The "D" in the DEGEST drivers of global change suggests that a chaotic future lies ahead.

Examining All Six Major Drivers and Their Implications for the Future

The purpose of this chapter is to explain why certain drivers of change are the most dangerous and explain the trend line dangers that these forces now represent. Identifying the drivers and monitoring their impact can set the stage for more

---

[3] Edward Cornish, *Futuring* (2004), World Future Society, Bethesda, MD.

proactive reforms. Only if we identify and monitor the most important drivers of change can we have a hope to reduce the threat to global civilization that these drivers now represent.

Currently, the tools we have to address these threats are too weak, and public understanding of the threats is too low to be able to craft the tools needed to combat these threats effectively. If we, as a species, plus the entire flora and fauna of our planet, are to survive mass extinction we need reforms – sooner rather than later. The first step is to understand the most important drivers of change and why they could threaten our very survival.

Let's briefly explore these forces, or "drivers" of change, in our twenty-first century world and understand at the outset that there are at least three "worlds" on our planet. There is the "first world" of the economically advanced countries of the OECD. There is the "second world" of the industrializing countries that are quickly evolving to be like the most economically advanced countries. Then there is the "third world," which represents the least developed countries. These countries are often constrained not only by poverty and struggling economies but also problems of political corruption, a lack of intellectual freedom, religious or societal interference in educational and social advancement, unsustainable population growth and perhaps even continued neo colonial exploitation. The problems of first, second and the third world countries are in many ways different, but we nevertheless all live on a single planet.

Ultimately, many of the most important problems, such as global overpopulation, climate change, unreformed economic systems and educational and health care problems, must and indeed do intersect. Scientists are now telling us we live in the Anthropocene age, where humans are the primary shapers of the key features of Planet Earth. The official definition of this new geological epoch according to the *Encyclopedia Britannica* is as follows:

> The Anthropocene Epoch is the unofficial interval of geologic time, making up the third worldwide division of the Quaternary Period (2.6 million years ago to the present). It is characterized as the time in which the collective activities of human beings (Homo sapiens) began to substantially alter the Earth's surface, atmosphere, oceans, and systems of nutrient cycling. A growing group of scientists argue that the Anthropocene Epoch should follow the Holocene Epoch (11,700 years ago to the present) and begin in the year 1950. The name Anthropocene is derived from Greek and means the "recent age of man."[4]

Too long we have pretended that what happens in Rwanda or Paraguay or Yemen or Indonesia has nothing to do with what happens in the United States or France or Japan. The world is a system – ecologically, economically, and politically. In today's world of global communications and transportation, we are interconnected in terms of natural resources, food and water, security, ecology and an increasing understanding of the urgency of finding a way to sustainably use the planet's finite resources. The planet continues to see our population grow exponentially and gobble up the resources that sustain us but which sustain many other species as well.

---

[4] John Rafferty, The Encyclopedia Britannica, May 2017, https://www.britannica.com/science/Anthropocene-Epoch

Fig. 3.1 A depiction of the world's water in relation to the planet – and only 1% is available for drinking. (Graphic courtesy of the Sierra Club)

The future of human population growth on Planet Earth will be driven by many factors, such as climate change, global warming, food supplies and more. Of all the factors, the one that might be most determining of our fate is a drinkable water supply. Figure 3.1 (prepared by the Sierra Club) shows that of the water supply on Earth, only 1% is currently available as drinkable water. The rest of the supply is salt water, polluted water, or non-accessible water. This is likely the limiting factor for human population growth. Today cities such as Bangalore, India, are facing this challenge as the millions of people from the Sahel region of Africa have been forced to relocate due to the lack of water.

It is some of the world's most economically advanced countries that, over the past two centuries, have given rise to many of today's global problems. When these powerful nations tell lesser developed nations that they are the ones that need to make the adjustments by such means as limiting population growth or not deploying low-cost coal-fired energy plants, they are naturally not overly receptive. They often respond that you created today's problems of greenhouse gases, climate change, resource depletion, pollution, and so on. Now you expect us to not follow your example? You are telling us to "walk the walk" that you never walked yourselves? What kind of justice is that?

Thus, in the chapters that follow, there is a need to address the drivers of global problems not only from the perspective of the most advanced economies but from the perspective of the least developed economies as well.

Population Growth and Super-Urbanization

For millions of years the world's nomadic human population remained below 100 million people, and then about 10,000 years ago viable agriculture suddenly evolved. Productive farming techniques quickly became possible when natural hybridization of wheat seeds suddenly allowed one seed to return three or four. Humans became fruitful and multiplied. In the seventeenth century the first of several "green revolutions" became possible, and seeds returned an even more bountiful harvest. Human population soared exponentially.

By the start of the nineteenth century the human population hit 800 million. As of the start of twentieth century this figure had soared to 1.8 billion. Then, at the start of the twenty-first century, the number was about 6.3 billion. By the start of the twenty-second century, the number may hit an astonishingly high number of 12 billion. If that hypothetical yet entirely possible world population of 2200 were to produce carbon-based greenhouse gases at a rate equivalent to today's typical American, that world might be hazy indeed. This population would be producing 3.6×10^{14} pounds of GHG per annum – almost a hundred times today's levels.

In the twelfth century, perhaps about 3% of the world's population lived in cities compared to about 53% of all people that live in urban areas, suburbs or exocities today (Fig. 3.2).

At the start of the fourteenth century Beijing, with a population of well under a million people, was the world's largest city. Geographically speaking, Beijing was relatively concentrated. Anyone in the heart of the city could go to its furthest extreme by taking a 5 km walk within its total area of about 50 km. It was a "pedestrian city," just like the rest of the world. Walking was the prime mode of transportation, and the prime source of energy was the burning of wood.

The current world is totally transformed in terms of population, urbanization, the size of cities, and the nature of almost everything from energy production to transportation and communications, and certainly human livelihood. More people and more urbanization change almost everything else. A world with 12 billion people aboard changes global economics, the ecology, governmental and social activity, and the rate at which technology evolves.

Climate Change and Unsustainable Growth

The problem with climate change is that what we do today can impact the world's ecology a hundred years – even a thousand years – from now. What we might change today might have modest costs, but it could have huge costs tomorrow. One recent

Growth of the City

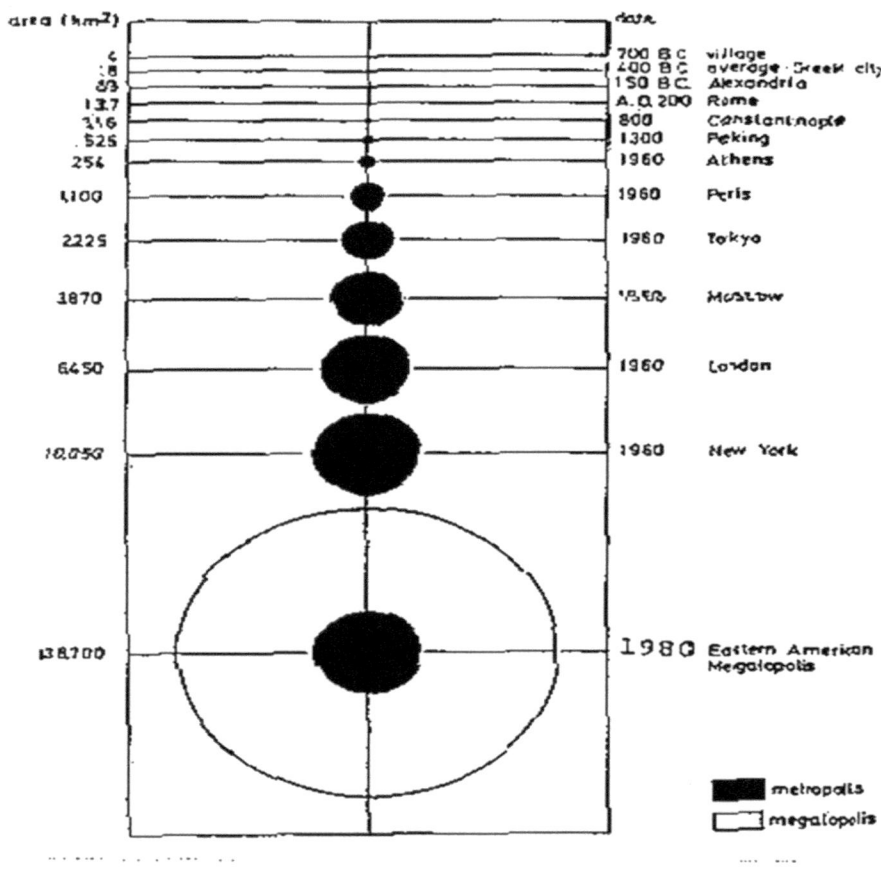

Fig. 3.2 The growth of the size of cities

estimate is that we can reverse a ton of carbon-based ghg for about $25–$35. But if we are producing something like 1.8×10^{11} tons of these pollutants a year that adds up to a rather large number of around $4–$6 trillion a year.

The population of humans on our planet increased by a factor of about three from 1800 to 1900. It increased almost another three times from 1900 to 2000. If we are lucky the population increase will be less than double in this century. The point is that there are many reasons why continued population growth cannot be sustained. Many doubt that even 12 billion people is a sustainable population.

In an age of smart machines, a threatened biosphere due to human pollution, and shrinking natural resources, the products of continued population expansion are all negative. The best way to mitigate climate change and global pollution is a smaller human population. The best way to mitigate technological unemployment is a smaller human population. The best way for all humans to enjoy a higher standard of living is a smaller human population. The best way to reduce international tensions and conflict over limited natural resources, arable land, and finite food and water supplies is a smaller human population.

All nations must embrace the goal of population reduction jointly if this is to succeed. Of necessity there will need to be both "carrots" to reward such action as well as "sticks" to create penalties for ignoring these goals. Singapore has used tax policy to limit growth. China and India have rewarded those who have exercised permanent means of birth control in a variety of ways. In developing economies rewards and penalties may need to be implemented at the village level. Delays in population controls will have longer-term economic, political and social penalties for everyone across the planet. Perhaps the most depressing fact of all is that human presence continues to have an adverse effect on more and more places.

We have polluted the land, the oceans, the arctic icecaps, and Earth's atmosphere and stratosphere. Our polluting practices have opened a hole in the ozone layer that protects us from radiation. In June 2010 the U. N. Committee on the Peaceful Uses of Outer Space formally initiated a study on the "sustainability of outer space."[5] Today over 27,000 pieces of space junk are being tracked in Earth orbit that are the size of a human fist. There are an estimated 500,000 pieces of space debris in low Earth orbit that are 1 cm in diameter. A piece of space junk with a diameter ten times less could kill an astronaut. In short, the presence of humans has not only fouled Earth but the space around our planet as well.[6]

Societal and Political Values of Un-moderated Economic Expansion

One of the enshrined goals of capitalist economies for at least the last three centuries has been rapid economic expansion based on ever-larger global markets. One of the premises of this growth has been to expand population to thus increase the size of markets and to accelerate throughput. The problem is that today both technology and the global ecology has changed the basic parameters related to these economic equations. Technology has eliminated the need for more workers to grow the economy. The polluting effects of more and more people are realized in a variety of ways that increase every day by such means as energy consumption, travel, heating and cooling of housing, recreation, etc.

---

[5] "Final Report of the United Nations Committee on the Peaceful Uses of Outer Space", Vienna, Austria, June, 2010.

[6] David Kushner, "The Future of Space Orbital Cleanup" *Popular Science*, July 2010, pp. 60–66.

Growth Versus Survival

In short, we need to maximize the value of sustainability and quality of life instead of expanding markets. Growth is not the objective any more. The goal is survival. The basic equations related to social, economic and political development have changed, and population expansion is now a liability rather than a positive benefit to society.

One "clean and green" smart machine that is available to work 24/7 is able to match the output of several people in terms of productivity. Such a machine does not require housing, transportation, food, water, education, health care or hospitals. Capitalism must be restructured to recognize new economic realities that include the high cost of pollution and the many non-sustainable aspects of twenty-first century human life styles – particularly in places such as the United States, Canada, Australia, Europe, and Japan. The U. N. Climate Change Conference in Copenhagen in December 2009 that merely asked countries to sign on to "an effort to keep the average temperature of world from going up by more than 2 degree Centigrade" was seen as an almost empty gesture. To date some 130 countries have signed onto the Copenhagen pledge, but this is a pledge that means virtually nothing unless backed up with action. Clearly today the stakes are much higher.[7]

The key to the future of a sustainable planet and the survival of people over the longer term is a modification of our economic systems to include in the cost pollution and practices that could eventually be destructive to life on Earth. We have already started some modest practices such as putting a high price on the cost of plastic bottles and plastic bags to encourage us to seek alternatives. Filters on water supplies, reusable grocery bags, and storing information electronically rather than printing it out on paper are some of the adjustments already made by simply applying the right economic incentives. The future needs much larger adjustments, such as legislation that will make it economically unsound to operate coal-based electricity plants. We may see "progressive" environmental taxes that will not only discourage wasteful energy, transportation or building practices but also give tax rebates to those that invest in "clean and green" autos or highly insulated homes and buildings. The biggest economic adjustment of all would be strong incentives to have no children or only one child. This may be the hardest economic reform of all to sell. Currently in some countries in Europe, where the population is stable or even shrinking there are incentives to have *more* children.

Super-Automation and Technological Unemployment

Most of the topics that are discussed in Toffler's *Fourth Wave* are known around the world as issues of concern and problems to be faced by twenty-first century society. The exception, however, is quite likely to be the social and economic implications

---

[7] "Interview with Yvo de Boer, Former Chief of the UN Climate Change Initiative," *Washington Post*, July 11, 2010, p. B3.

of a rapidly evolving super-automation. Particularly in Korea and Japan artificially intelligent automation and "knobots" are evolving rapidly. At research labs, such as The Korea Advanced Institute of Science and Technology in Korea, the Institute for Physical and Chemical Research in Japan, as well as the Intel Labs in Pittsburgh working with researchers at the Carnegie-Mellon University the world of increasingly smart and capable robots continues apace. Investment in this technology is currently led by Japan and Korea, followed by Europe. China is increasingly entering this field. The investment by the United States in this research lags behind most of the other advanced countries.

However, the near-term wave of technological unemployment is not about smart robots. The key in the near term will be expert systems and AI software that can be loaded into computers and produce answers that people once derived. Today one can go to Quicken and buy a program to complete income tax forms to a degree of sophistication that required a CPA in the past. Accounting programs, real estate assessment programs, interactive training programs, and art and ad layout programs are replacing people and professions at a rapid rate. Even medical diagnostic programs are replacing nurses, nurse practitioners and some doctors. Over the next few decades more and more service jobs will be replaced, and the pace of technological unemployment or underemployment will only increase.[8]

In some ways this will be good. The rapid aging of the population within OECD countries will require more and more assistance for the elderly. Smart machines will aid this special population and its needs and make the aged less a burden on society. There is today a wide range of robots – some of the reasonably smart – that have been designed to care just for the aged, the physically infirm or the medically needy population.[9] In the longer term, however, only a reduced population will help a true employment crisis as the transition triggered by super-automation continues apace over the next 30 years.

Unreformed Health Care and Educational Systems

In 2010 a renewed cry went up in America to address the U. S. longer term fiscal crisis that would involve a wide range of medical and social benefits for the elderly at a time when a huge number of baby boomers would be retiring from the work force and drawing out cash from an underfunded Social Security system.

When Congress was not able to pass legislation to set up a bi-partisan commission to address this issue, President Obama proceeded to establish such a commission headed, by former Republican Senator Alan Simpson from Wyoming and former Chief of Staff during the Clinton Administration Erskine Bowles. The objective of the 14-member commission was to bring revenues and expenditures of the federal

---

[8] Joseph N. Pelton and Peter Marshall, *Megacrunch: Ten Survival Strategies for the 21st Century*, (2010) PA Associates, United Kingdom.

[9] Corey Binns, "Rise of the Helpful Machines", *Popular Science*, August 2010, Vol. 277, Issue No. 2, pp. 44–51.

budget into balance at no more than 21% of the Gross National Product.[10] This objective was never achieved during the budgetary processes that defined the two terms of President Obama, and do not seem likely to be achieved within the term of President Trump.

However, the significant fact is that nation after nation from Spain to Greece, from Eastern Europe to across Africa and South and Central America, the problem of budget deficits has spread. More and more governments have recognized that business as usual cannot continue without the worldwide spread of bankrupt governments. In the wake of the largest global recession since the Great Depression of the 1930s, which began in 2008 and lasted for at least 6 years, there is an ever more widespread recognition that many of the social policies must be rethought. With a population where 80-year-olds are growing apace, universal retirement at age 55, such as in France, or federal or teacher retirement after 30 years of service such as in the United States, is simply not viable. Virtually free educational and medical services for an ever-growing percentage of the population without dedicated tax revenues is not economically feasible in either a capitalist or socialist society. If people live longer then older retirement ages must ensue.

If medical care costs escalate, then either one must find a way to cut costs, reduce benefits or find a way for supplemental retirement and medical insurance to be funded, either individually or on a group basis. Most significantly it seems likely that new technology in both the fields of education and health care can, in time, make these services more efficient, more proficient, and cost effective. IBM's development of its prodigious "Watson" super computer, loaded with exabytes of information, and especially its Watson system devoted medical information, is just one of the new tools that is being developed to support tele-health-based services.

The United States, with dubious distinction, leads the world in its spending on education and health care – but with the least productive results. Within the United States at least 10% of its GDP spending is on education, not including extensive training programs, and about 16–18% of GDP goes to spending on health care. The insult to injury aspect of this statistic is that the results are clearly worse than in nations that spend far less of their money on these services. The students from many of the OECD countries score higher on standardized tests than the United States while spending less than what is spent in America.

Likewise, there are many countries where their citizens live longer and where there are less childbirth deaths than in the United States, despite the level of spending being far more. The following charts suggest that the United States has a lot to learn from other countries in both fields of education and health care. These charts also indirectly suggest the wisdom of developing and applying better and "smarter" technology in addition to carrying out legislative and structural reforms to revamp American health care and education. The alternative is to see an increasing percentage of students and elderly suffering from substandard practices.

---

[10] David Broder, "Glimmers of Hope on the Budget Crisis," *Washington Post*, July 15, 2010, P. A19.

Table 3.1 OECD health care data on key health care statistics

Country	Life expectancy	Health care costs/cap ($ US)	Health care costs as % of GDP	% of gov't funds going to health care	% of total health care paid by gov't
Australia	81.4	3137	8.7	17.7	67.7
Canada	80.7	3895	10.1	16.7	69.8
France	81.0	3601	11.0	14.2	79.0
Germany	79.8	3588	10.4	17.6	76.9
Japan	82.6	2581	8.1	16.8	81.3
Norway	80.0	5910	9.0	17.9	83.6
Sweden	81.0	3323	9.1	13.6	81.7
United Kingdom	79.1	2992	8.4	15.8	81.7
USA	**78.1**	**7290**	**16.0**	**18.5**	**45.4**

OECD Health Care Data 2010, http://www.oecd.org/dataoecd/

Table 3.1 shows that U. S. life expectancy falls below eight other OECD countries, even though its health care costs are sometimes double that of these other countries and that health care consumes a larger share of the U. S. Gross Domestic Product. The U. S. federal government also spends a higher total of its national budget on health care (i.e., 18.5%) even though this money represents only about 45% of the total cost of health care. These figures adds up to what could be considered a "lose, lose, lose" situation for the United States.

If one suggests that longevity is not a fair measure of health care, one could also opt for infant mortality rates. Here, too, the United States lags behind these other countries with a mortality rate of 6.5 per thousand compared to about 5–6 per thousand for these other countries.

Educational statistics as complied by the OECD and other sources again shows that the United States spends more on its educational programs but with less than stellar results. Table 3.2 above shows that the United States lags far behind other countries when it comes to comparative test results. Although Table 3.2 shows results in math, science test results also show a similar pattern. Among OECD countries only the Republic of Korea and Iceland spend on a per capita basis a comparable amount of its GDP on education to the United States.

The United States certainly in absolute terms outspends all countries. Even in terms of a percentage of its GDP, the United States outspends, in descending order: Denmark, Canada, Sweden, New Zealand, Belgium, Slovenia, France, Switzerland, the United Kingdom, Finland, Mexico, Australia, Poland, Chile, Hungary the Netherlands, Portugal, Austria, Norway, Japan, Italy, Brazil, Czech Republic, Germany, Ireland, Spain, the Slovak Republic, the Russian Federation and Turkey.[11]

The future trend lines for health care and education show are disturbing. People are living longer and thus need health care even past 100 years of age. Medical

---

[11] "Education at a Glance: OECD Indicators: Chapter B: Financial and Human Resources Invested in Education," http://ww.oecd.org/dataoecd/

Table 3.2   U. S. student math scores compared to other nations, Grades 4, 8 and 12

	Grade 4		Grade 8		Grade 12	
Rank	Nation	Score	Nation	Score	Nation	Score
1.	Singapore	625	Singapore	643	Netherlands	560
2.	Korea	611	Korea	607	Sweden	552
3.	Japan	597	Japan	605	Denmark	547
4.	Hong Kong	587	Hong Kong	588	Switzerland	540
5.	Netherlands	577	Belgium	565	Iceland	534
6.	Czech Republic	567	Czech Republic	564	Norway	528
7.	Austria	559	Slovak Republic	547	France	523
8.	Slovenia	552	Switzerland	545	New Zealand	522
9.	Ireland	550	Netherlands	541	Australia	522
10.	Hungary	548	Slovenia	541	Canada	519
11.	Australia	546	Bulgaria	540	Austria	518
12.	United States	545	Austria	539	Slovenia	512
13.	Canada	532	France	538	Germany	495
14.	Israel	531	Hungary	537	Hungary	483
15.	Latvia	525	Russian Fed.	535	Italy	476
16.	Scotland	520	Australia	530	Russian Fed.	471
17.	England	513	Ireland	527	Lithuania	469
18.	Cyprus	502	Canada	527	Czech Republic	466
19.	Norway	502	Belgium	526	United States	461
20.	New Zealand	499	Sweden	519	Cyprus	446
21.	Greece	492	Thailand	522	South Africa	356
22.	Thailand	490	Israel	522		
23.	Portugal	475	Germany	509		
24.	Iceland	474	New Zealand	508		
25.	Iran	429	England	506		
26.	Kuwait	400	Norway	503		
27.			Denmark	502		
28.			United States	500		

"Math scores Academic Failure-International Test Score Results," http://4brevard.com/choice/international-test-scores.htm. A similar pattern can be seen in science test scores with American students slipping in performance against students in other countries as the progress from grades 4 to 12

research is expanding, and the amount of knowledge doctors, nurse practitioners, and nurses must acquire is daunting. The cost of medical care and drugs is outstripping inflation. The number of doctors per capita is shrinking, especially in rural and remote areas. Education has similarly adverse trend lines.

The number of people to be educated in the twenty-first century is larger than that represented by all the people requiring education since ancient times, when formal teaching began. The amount of new information being added the global information database is now expanding by at least 4 exabytes a year. That is 4,000,000,000,000,000,000 bytes of information (or the equivalent of several quintillion words). This plethora of information added each year is equivalent to many millions of times the amount of information stored in ancient Greek, Roman, Middle Eastern and Chinese societies. The amount of information that teachers and students are asked to know and assimilate has mushroomed in just the last century and continues to expand exponentially. Again, in contrast, the number of teachers per capita is decreasing.

The bottom line is that expert systems, artificial intelligence, the Internet of Everything, and eventually self-aware machines will need to assume a greater and greater role in both health care and education to meet the demands of a twenty-first century world. Computer-assisted education, self-directed education using video labs, and computerized learning systems are likely to be a significant part of the future of education. Likewise, these techniques will need to be applied to the training and education of doctors, diagnosticians, nurse practitioners, and nurses. Already there are sophisticated tele-health programs that deliver increasingly high quality health care to rural and remote areas, and the trend will undoubtedly continue. Part of the problem is in information systems that process and store information inefficiently and inefficiencies in health care insurance companies and governmental support agencies. Other problems arise between the confusion that exists in many countries between what is, in fact, "child tending" for working people and what is education.

What is the future of health care and education in the United States and OECD countries, and what is the future of health care and education for the rest of the world? The answer is far from clear, but one thing that is sure. Almost everything will be different. These services will be different in terms how they are delivered. They will be different in terms of who delivers these services and how either people or smart machines are trained or designed to perform. They will be different in terms of cost structure, governmental agency organization, and industries that support these service industries. Finally these systems must also be geared to teaching young people of the value and urgency of converting to a circular or sustainable or Fourth Wave economy. Virtually all new economic enterprises and governmental activities must be geared to creating new enterprises and jobs focused on green energy systems, control of pollution and curtailing of population growth.

As Abraham Maslov has explained in some detail in his writings and his research, people do not aspire to meeting their higher-level needs until after their basic biological and safety needs are met.[12] This then becomes the great challenge. One

---

[12] Abraham Maslow, *Motivation and Personality,* (1954) Harper and Brothers, N.Y.

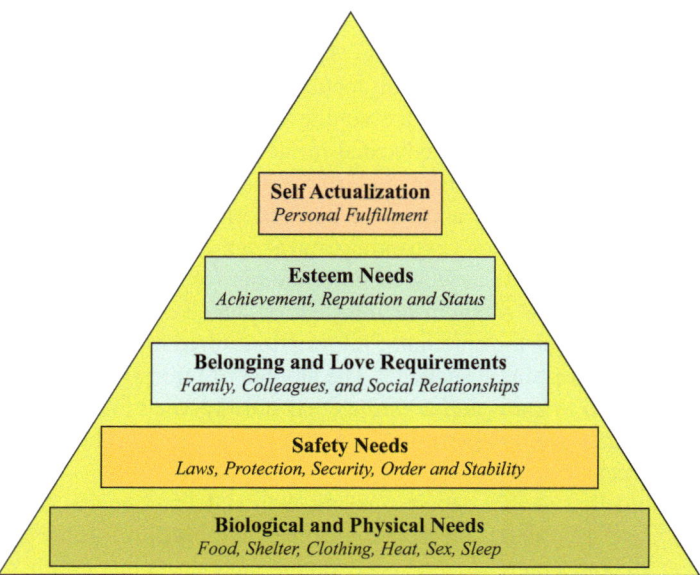

Fig. 3.3 Maslov's hierarchy of human needs. (Graphic by the author)

must thus find creative strategies in combining ways to meet biological, physical, and safety needs with the creation of new jobs and enterprises designed to also meet goals for sustainability, clean energy, population moderation, and a circular rather than a disposable economy. Space 2.0 and Cyberspace 2.0 must be enlisted at every level to make these sustainability goals a top priority (Fig. 3.3).

Life in a Broadband Cybernetic World

The force of future compression in terms of ever-accelerating rates of social, technical, and cultural change is seen most significantly within the world of telecommunications, information technology, networking, and artificial intelligence. Within a period of just over three decades people have gone from the advent of personal computers and cell phones to a world where we are inundated by social media, the Internet of Things, smart robots, online expert systems, networking trolls, the "dark web" and nations orchestrating cyberattacks on other nations and interfering in national elections, such as Russia did in the U. S. 2016 presidential elections. And this electronic technology and artificial intelligence is moving ahead at light speed. Researchers anticipate being able to create the equivalent of an electronic human brain within a decade and to design and sell at reasonable cost the equivalent of a self-aware machine with human intelligence and mobility within two to three decades. Our political, legal, social, economic and technical systems are not geared to adjust to such rapid change, and the consequences are enormous.

Coping with the Future

Today we are seeing impulse jumps in hacker attacks and cybercrimes to steal money and assets as well as to stage assaults on institutions and even national governments. We are seeing automation becoming a much larger threat to employment than the movement of jobs offshore to lower cost labor markets. Smart machines will bring home manufacturing from overseas, but these jobs will go to super-automation and robotic production and not to unemployed laborers. Broadband telecommunications, mobile electronic services, networking, automated systems, robotic manufacturing and services will create job loss around the world. Under-employment and loss of service jobs will occur at an impressive and even alarming rate. Future compression will be on a tear as these disruptions of employment rates accelerate in pace.

The average citizen will feel more threatened by the technology instead of enabled by it. The loss of jobs, meaningful employment, and the ability to assimilate ever advancing technology will create more angst than opportunity – unless?

Unless political, social, cultural and even technological change is undertaken, along with the other proposed innovations related to birth control and climate change, to cope with this brave new world. Technological innovation is speeding up ever faster, but the various forms of legal, regulatory, political, social, and cultural processes needed to adapt to these changes are increasingly out of synch with this change. For instance, it is hard to think of a job that one can identify today that will not be performed with higher competency and lower cost and over longer hours of operation by smart machines. We are simply not ready for this world.

We must re-invent the future by recognizing the importance of combining economic growth, sustainability, and coping with climate change and population growth. This needs to permeate considerations such as tax policy, governmental public works and services, job creation, cleaner ways to create and use energy, transportation systems, educational and health care systems, and every aspect of modern life and economic and social activity. In the next chapter the specifics of the Space 2.0 and Cyber 2.0 revolution will be explored.

Chapter 4
Smart Cities, Megacities and Meta-Cities

A city is defined by its sense of community.

Paul and Percival Goodman

Introduction

The world today has about 30 cities with a population of about 10 million or more residents. These megacities are larger in size than many countries. They act as magnets to people from rural areas in search of jobs and employment that pay better wages than they can command in farming and remote areas. Today, the world for the first time is more urban than rural.

For millions of years humanoids lived in nomadic groups or tribes that were hunters and gatherers, and there were no towns or cities. It was only 10,000 years ago that humans invented farming and towns began to appear. It took a long time for cities to appear. Even 300 years ago the world was still predominantly rural. Perhaps only 3% was what might be called urban and 97% rural. But since that time towns and cities have grown and grown. It was only at the start of the twenty-first century that city life grew to be the predominant way for people to exist. Now, however, times are changing – and rapidly so. It is expected that by 2100 the world will be 80–85% urban. This means that billions and billions of people are expected to pour into cities. By the year 2100 there may be over 100 megacities of over 10 million.

There are projections that say that by 2050 there will be an additional 1.5 billion people or more born, while the total increase in urban population by 2050 may be as large as 2.5 billion. This means that the magnetic pull of cities around the world will greatly increase the mass migration of people to them. In two books by the author, co-written with Dr. Indu Singh, namely *The Safe City* and *The twenty-first*

The original version of this chapter was revised. The correction to this chapter is available at
https://doi.org/10.1007/978-3-030-02137-5_11

© Springer Nature Switzerland AG 2019
J. N. Pelton, *Preparing for the Next Cyber Revolution*,
https://doi.org/10.1007/978-3-030-02137-5_4

Century Smart City, we suggested that there is a sort of an axiom of logical development when it comes to cities. This axiom, or 'law,' is that sprawling and highly distributed settlements around cities tend to be ineffective and economically inefficient. Sprawling settlements are, in effect, undesirable, while urban density is good. But this rule of effective density can reach a point of diminishing returns. Super mega density can become increasingly inefficient. When city populations soar into the many millions, there are increasing inefficiencies. These inefficiencies increase with high rise buildings greater than 50 stories, and a hugely congested urban core tends to create problems.

These problems, that can stem from too much density, include: (i) excessive air and water pollution that can become a health hazard; (ii) limitations on the part of first responders in the event of disasters and/or terrorist attacks as well as inefficiencies in terms of policing, fire-fighting and medical emergencies, particularly in super high skyscrapers; (iii) inefficiency of transportation systems and massive tie-ups in daily commuting; (iv) distorted real estate values for both housing and offices.

The key questions here are what should city planners do if they are faced with rapid and indeed surging urbanization, massive overcrowding and excessive new demands on their infrastructure? Can they make their services and facilities more efficient and 'smarter'? Are there better planning mechanisms that can emerge from big data analysis? Is it possible to relieve the pressure on megacities by diverting new development and overcrowding within their massive intra-city cores to satellite 'meta-cities' in order to relieve transportation and housing congestion? Do smart cities help make cities more livable and viable by such means as encouraging tele-commuting?

The Coming Emphasis on Smart City Technology and Systems

There is an almost endless list of questions as to what innovations associated with smart city planning and intelligent infrastructure can be deployed and used to make cities better. Areas of focus include making cities safer and more economically viable with an enhanced tax base. New technology can allow smart cities to be better equipped to deliver health care and education services, be more responsive to resident and business interest needs, become more cost efficient, and to contend better with environmental concerns, global warming, overpopulation and intense crowding.[1] (See Fig. 4.1).

Cities in various parts of the world will interpret and implement smart city technology and systems in quite different ways. The differences hinge on many factors. There is first the difference between states with strong governmental controls versus those with strong democratic traditions that may place many restrictions on technology related to citizen's rights to privacy, including restrictions with regard to surveillance. There is also the difference between countries at

---

[1] See presentation by Joseph Pelton and Indu Singh, Digital Destiny Series of Public Forums, Arlington, VA.

BENEFITS
- Better planning and economic growth
- Less pollution and reduced energy use
- Better and more resilient infrastructure
- Less commuting and better transport systems
- Reduced traffic accidents
- Better crime control
- Responsive government action
- Cost-efficiencies and Competitiveness
- More viable tax base

CONCERNS
- Cyber attacks
- Techno-terrorists
- Vulnerable industrial controls
- Human-machine interface weaknesses
- Maintenance and updating
- Reduced human Interaction
- Reduced privacy
- Trolling and threats to democratic processes

Fig. 4.1 The pros and cons of smart city planning, technology and systems. (Permission provided by the author. All rights reserved by J. Pelton and Indu Singh)

different levels of economic development that have different stores of financial reserves to invest in modernization and new infrastructure and control systems. Appendix 2 in this book contains the groundbreaking European General Data Privacy Regulations (GDPR) that is now having a worldwide impact. Appendix 3 outlines the U. S. efforts to draft a law to create a consumer data privacy Bill of Rights that has so far not managed to be passed into law.

There are also concerns with regard to employment, and thus decisions with regard to automation and autonomous control may also be impacted by the need to retain full employment. There are, of course, other dimensions of smart city planning that relate to concerns for cybersecurity, potential terrorist attacks, natural disasters, emergency services, and community safety.

The bottom line is that each city will need to consider the stew of ingredients and systems that makes the most sense for its needs and longer-term vision. Each city and its political leaders must consider what new technology, systems, planning techniques and public input and review processes will be employed to obtain public support and endorsement of these changes.

Nevertheless, despite this great diversity of thought and differences in approach that various cities around the world will take in moving toward smart infrastructure and controls, virtually all cities will face the revolution that is coming. The entire world is moving toward a higher level of automation, adopting more technical sophistication and seeking resiliency against natural disaster, cyberattacks and terrorist efforts to disrupt a city's day to day smooth functioning.

How Urban Life and Especially Smart Cities Will Be Transformed

What is here being called "the revolution" is, in fact, recognition of a dramatic transformation that is coming to the world and especially to urban life – perhaps in the next few decades. This revolution includes super-automation and the transformation of jobs and work (this is to say, the Fourth Wave economy). It includes reforms and changes in laws, regulations, and human practices to adapt to the new world that is coming. This new world will be characterized not only by a Fourth Wave economy but also massive amounts of population growth and overcrowding in cities and the demands imposed by climate change – particularly in seacoast cities where literally billions of people live. To many people smart cities means something like the use of more technology, more automation, and modern infrastructure. In Fig. 4.1 the key aspects of a smart city are defined in terms of better planning and citizen participation in that planning, better safety, more effective environmental controls and a better response to climate change, etc. In short, a smart city is one that can use technology to provide better and cheaper services to citizens and businesses. In some cases this is new and better technology. In other words it is better planning and decision-making processes, more effective understanding of what is working and succeeding by using big data analysis of past performance.

Transportation in the Smart City

Some believe that the best way to improve transportation is to build freeways with more and more lanes of throughput. But we have been there and done that. Part of the revolution is to look at new and perhaps disruptive ways in planning for transportation in the twenty-first century. This could change thinking in several ways. One new model might be based on far more telecommuting to work and thus to move ideas and communications about a city more than people. Another idea is to create modernized meta-cities around the periphery of megacities; this changes the broad architecture of a city to be more like a chessboard than a target with super density high rises at the internal core of a large city, where congestion, traffic snarls, goliath buildings and angry tempers can flare.

This super density at the core of megacities, with its high rises that soar into the sky, can create enormous problems. A 100-story super building is impressive, but when it comes to responding to fire, crime, terrorist attack, earthquakes or other disasters such super density becomes a problem. The exponential rise in air pollution, the cost and congestion of traffic systems, and the inordinately high cost of land at the core become prohibitive.

NEC, which is headquartered in Tokyo, has moved towards creating over a 100 telecommuting locations for its workers. This was perhaps a matter of convenience and becoming aware that a good portion of its workers were spending up to 4 h a day getting to and from work. The main driver, though, was in recognizing that the costs of creating satellite telecommuting/telework centers was a small fraction of the cost of building a new high-rise in Shinjuku, where its headquarters are located. There was one study done on what would be the value of developing the land represented by the Imperial palace grounds in Tokyo to the level of density represented by the Ginza and Shinjuku areas of Tokyo, and the answer was that it would be equal to the value of all high rises in Canada.[2]

Smart transportation can indeed see the development of new technologies such as Maglev or hyperloop train systems, hypersonic planes or driverless cars and trucks, and such innovations will happen. Nevertheless, it is possible that the most important innovations will be to find ways so that people will need to move less. Using telework centers not only relieve traffic congestion but also greatly serves to reduce air pollution. In the smart city of tomorrow the linkages between pollution, climate change, transportation, IT and communications systems and urban architecture will increasingly be seen and recognized in planning for the future.

Energy

It turns out that in most cities about 70% of all energy consumption comes from heating, cooling and operating buildings, and about 20–25% relates to transportation systems. It is important to apply intelligence and smart design to buildings and transportation systems to ensure that much less energy is consumed, so as to lower costs and pollution. If office and retail buildings, hotels and homes are engineered with better insulation and are designed or retrofitted use sustainable energy sources, the costs of operation can be reduced as well as pollution from greenhouse gases. Further, if our cities are served by more energy efficient transportation systems, such as electric cars, trucks and buses that no longer use hydrocarbon fuels, this will not only make our cities cleaner and less polluted but in the longer term the costs of energy consumption will also come down. In the near term the costs are generally higher, but by around 2030 costs will fall and continue to fall as the technology associated with photovoltaic cells, electric motors for vehicles and other sustainable energy systems are perfected.

The largest and most difficult parts of this transition will be in energy systems and energy use. In this regard, large oil companies and operators of electric energy systems will typically resist these changes, because they have billions of dollars invested in old technology such as coal-fired power plants, oil refineries, gasoline service stations, and more. The greatest challenge may be in finding ways to move the energy companies of various types (electric power plants, nuclear energy plants, oil refineries, etc.) from old-fashioned, polluting, dangerous and outmoded technologies into new technologies and sustainable practices so that they can find a favorable and profitable transition to the future. Because of their strong lobbies, they

---

[2] Joseph N. Pelton "Designing the Future City" in Indu Singh and Joseph N. Pelton *Future Cities*.

can put pressure on politicians to preserve energy technologies that fuel pollution, accelerate climate change, and ultimately serve to increase the cost of energy to consumers. This is one of the big transitions of the revolution and one of the areas where there can be a win-win for consumers and business – if smart planners in cities can find the right path forward.

Environmental Systems and Climate Change Concerns

The big changes that can come in transportation and energy in the next few decades can also make a huge difference with regard to environmental systems, pollution and climate change concerns. The focus of smart cities in dealing with environmental concerns has broadened over time to include many features. One important area is that of recycling. There are many advantages to recycling and creating a circular economy. These advantages typically can include:

- The use of less energy rather than mining and fabricating from new raw resources.
- Less pollution, particularly with regard to plastics. By using less energy, a more circular economy can also result in less air pollution.
- It can save natural resources by reducing the need to mine metals, and by producing less plastics, glass, etc.
- Saving scarce landfill space, which is a large problem for megacities, and again saves energy if disposal sites are located far away from the city.
- Recycling, which has become an industry and typically creates extra income and jobs. It also typically helps to reduce littering and burning of trash that again helps with the reduction of air pollution.

Recycling programs have expanded to include paper, cardboard, plastic, glass and metals that are often processed at a single cart without sorting. Leaves, grass, and branches are collected and processed into leaf mulch and bark mulch. The next frontiers in the move toward 100% recycling are with regard to creating composting systems for organics and unused food, systems that can recycle old appliances such as refrigerators and washing machines into metal chips and the recycling of Styrofoam. The recent discovery of an enzyme that can eat and break down plastics seems to be one of the more important discoveries with regard to recycling plastics.[3]

The prime focus of smart city environmental programs is typically to reduce air pollution, move to a zero carbon footprint and provide a prime way to combat climate change and the increase of global warming, with the attending rise of seawater levels. One of the most common means of measuring progress with regard

---

[3] Joah Gabbatiss, Plastic-eating enzyme accidentally created by scientists could help solve pollution crisis, IndependentNews, U.K. April 16, 2018. https://www.independent.co.uk/news/science/plastic-eating-enzyme-pollution-solution-waste-bottles-bacteria-portsmouth-a8307371.html

to reduced pollution is to measure reduced energy consumption within buildings and in transportation systems – particularly with regard to car, truck and bus traffic. Within the smart city community, the close coordination of energy, transportation, and environmental and climate change programs has become increasingly important. Some of the key steps that have been taken by forward-looking cities are master plans for energy, transportation and environment and climate change. The key to such master plans is the developing of metrics that can be monitored and progress measured to understand trend lines.

Telecommunications and IT Systems and Government Services

Another area that has become linked to the other sectors of energy, transportation and environment is that of telecommunications and IT systems. It has been recognized that telecommunications, broadband communications and advanced information technology (IT) are tools that can improve the efficiency of government services in a myriad of ways and that telecommunications and IT can be used as an alternative to transportation and to save the use of energy. Again a number of smart city administrators are requiring the development of master plans for telecommunications and IT systems.

The ways that broadband digital systems as well as artificial intelligence and smart robotics can improve the efficiency of city operations is almost endless. Broadband digital networks and mobile communications systems can allow city building inspectors to do more inspections and file reports on smartphones, notebooks and laptops rather than returning to the office. Police, fire and emergency medical treatment professionals can improve their performance in countless ways in terms of receiving emergency alerts, getting information quickly to remote sites and obtaining updates from sensors or even weather or remote-sensing satellites in near real-time conditions. Satellite phones are now nearly essential equipment in responding to emergency disaster conditions such as hurricanes and typhoons that have caused electric power to fail.

These devices are also a potential boon to citizens and businesses in terms of knowing buses or trains schedules and delays, when public hearings are being held, or to remotely attend public boards, councils or commissions as these events are streamed online. Some cities have installed ultra-high speed fiber optic networks that can operate at speeds up to terabits/second to support public governmental services, public education lectures, remote experiments, tutorials for teachers and video conferencing among teachers or administrators. These can be of particular benefit during emergency or disaster conditions, including snowstorms or other inclement weather, thus allowing workers to telecommute to work. These tools can allow automation of many governmental tasks, remote storage and back up of vital records and thousands of other tasks that can be accomplished faster, more efficiently and with less energy and office space.

Such urban communications and IT systems can also allow local government to offer business or other governmental units expanded capacity at low cost to reach other entities that need such services. It also allows for the use of master teachers to teach special classes among different schools, as well as to stream board and commission meetings to residents on the local cable television stations.

The increased use of telecommuting has also led to so-called 'hoteling' with regard to office space and thus reduces the amount of office space needed to be paid for and maintained. Dispatch of work crews is also more efficient.

These smart information systems can provide greater security and resilience against natural disasters. Traffic signals can be controlled to allow an emergency evacuation whether for a hurricane, tornado, flood, or terrorist attack. The emergency 911 center can be equipped to handle telephone, broadband cellular and text messaging. It can also be equipped to link to other police and fire jurisdictions to deploy vehicles from adjacent cities if their facilities are closer at hand. The smarter facilities and infrastructure are, the more flexible, responsive, and cost effective they can be.

One of the recent innovations is the remote control of street and pedestrian lighting that can be modified during the course of evening, night-time and early morning hours in order to save electricity. In short, broadband digital services can make a community smarter in many, many different ways.

Utilities

Although electrical energy and power efficiency is in many ways the top concern within a smart city in terms of continuity of urban services, there are other vital utilities to be addressed. All utilities must be reviewed in terms of continuity of service, reliability, resilience, and response to natural disaster and potential terrorist attack. Water and sewerage services are vital, and the latest technology is needed to keep water pure, sewage treatment at state-of-the-art levels, and safe from mishaps or problems with supervisory control and power failure, pipeline control failures or some sort of natural disaster or terrorist attack – either direct or via cyberattack. All communities to be considered smart should have an effective means to monitor the security of SCADA (supervisory control and data acquisition) systems and to provide sufficient security to their water, sewerage and utility control systems.

Although natural gas, electrical power, telephone and perhaps other utility systems may be owned and operated by private companies, urban governmental systems should have a strong and effective means of monitoring the operation of these various systems to make sure that their control systems and network distribution systems are up-to-date with adequate security controls and ability to respond to emergencies or potential cyberattack.

Education and Health Care

Yet another vital area for a smart city that will become of increasing importance in the coming revolution will be in the area of education and health care. Today education and health care tend in many ways to be standardized, with a one-size-fits-all type of mentality. Students learn in classrooms as if they have exactly the same abilities, talents and interests. Health care is provided in much the same standardized way, with annual checkups and blood, stool and imaging tests that are geared to age groups or ethnic background or sex. In coming years we will begin to see the sort of customization that futurists such as Alvin Toffler spoke of in *Future Shock*.[4]

In the age of broadband networks, AI, software systems that are today represented by IBM's Watson, smart robots, and eventually the Singularity, the world will move to more and more specialization geared to the individual capabilities and educational and health needs of individuals. DNA testing and genome analytics will, beginning at birth, indicate a good deal about future intellectual capabilities, athletic talents and medical needs as well as potential health problems of individuals.[5]

Educational and health care programs can be much more geared to the capabilities, genetic frailties and opportunities that are largely written in one's DNA. This does not mean transitioning to an overly structured and mandated future, such as was foretold in books like *Walden Two* by B. F. Skinner.[6] Yet the latest information technologies and breakthroughs in cyberspace technologies and twenty-first century research can allow medical problems to be diagnosed and treated much earlier. It means that the Albert Einsteins, the Isaac Newtons, the Maya Angelous, the Robert Frosts and the Pablo Picassos of the future might be identified and nurtured from an early age.

These opportunities and also associated problems are yet to be considered in detail. In coming chapters there will be further consideration about the opportunities, as well as potential pitfalls, that can be anticipated in the area of education, training, health, and medical care during the age of the revolution.

Smart Urban Planning Systems

The key to smart cities will be smart planning systems. Not every new technology is good. Not every new toy dreamed up by an inventor, engineer or scientist will advance the human destiny. For towns, cities, regions, nations and the world of the

---

[4] Alvin Toffler, *Future Shock.*

[5] "Cognitive Healthcare Solutions, IBM Watson Health, April 2018, https://www.ibm.com/watson/health/

[6] B.F. Skinner, *Walden Two.*

future, the key to a better future is smart planning. And here space planning is defined as being based on a common vision of the people within a community as to what they aspire to be. City planners should not base their decisions to buy a new 'gee whiz' transit system, or a water processing plant, or a solar energy system for public buildings on what a sales rep tells them. No, they should have long-term plans for their community in terms of overall goals for energy, transportation, utilities, education and health care, cultural opportunity and amenities, parks and recreation, tax base viability. Business expansion and development and effective government. They need a vision as to how it works together and how it evolves over time to meet the needs of changing demographics as residents age, new business opportunities emerge and older businesses fade. There must be an organic view of the holistic future. The main thing is to remember what the horse is and what the cart is. The visionary and holistic plan should be the 'horse' that drives the urban future and technological innovation should remain the 'cart' that follows the horse's lead. Technological adventurism can block or confound future vision as often as it will enable it.

Security, Resiliency, Cybersecurity and First Responders

The prime key to a safer and better future for cities, despite their size, may lie with well-conceived security plans for a better urban future. The key to security planning is to consider all the types of threats to safety and create a comprehensive plan that takes all future dangers into account. Dangers to future cities are manifold. They range from hyper-object types of dangers, such as climate change, global warming and pollution, excessive population and overcrowding that can overwhelm city infrastructure and services, and massive under-employment and joblessness that could come with a Fourth Wave economy. Here cities need to heed the axiom of thinking globally and acting locally. Cities around the world, for instance, are taking on the challenge of seeking to achieve a zero carbon footprint and also acting to cope with issues that are global in scope within their own communities.

There are other dangers that must be faced locally. These include possible natural disasters and how to create a city with buildings and homes that are designed to cope with fires, earthquakes, droughts, floods and more. Mega-skyscrapers that are above 50 stories high are especially difficult to defend against either natural disasters or terrorist attack. In fact, megacities themselves are very hard for first responders to cope with and maintain safety in light of a large-scale disaster, whether from natural or human-initiated attacks such as a terrorist attacking a water supply, exploding a bomb or flying a jet into a building.

Cybersecurity, in particular, is increasingly a threat to cities. City-states such as Singapore and Dubai have now moved to employ block chain security for vital records and vital infrastructures in order to protect against terrorist attack. The caution here is that when you have computer-controlled systems with humans in the loop, there is always the danger of an insider breach of security. This means that machine-driven systems must always have a potential brake on them to guard against malware, cyberattack, and insider security leaks of vital information and protection codes. In effect, all cyber systems must have functioning human-machine interfaces that can shut down an automated system that is under attack. These issues are of particular and mounting concerns in the coming years.

Conclusions

In future decades, the world population may swell to unprecedented numbers. More people will need to be educated and receive health care. During this time the percentage of humans that live in cities may expand to 80%, 85% or even 90%. The future of cities is, to an extent, where the future humans will increasingly live. These cities will, unfortunately, be more and more vulnerable, and megacities of 10 million or more will be the most vulnerable of all due to reliance on consolidated and massive energy grids, transportation systems, water supplies and sewage systems, utilities and other infrastructure that might be lost due to natural disasters, physical assaults by terrorists or the mentally ill, or by cyberattack.

The twenty-first century offers a wide range of opportunities for humans that live in cities to obtain better education and training, better health care, more access to all the amenities that urban centers can offer in terms of parks and recreation, sports and amusements, cultural attractions, libraries, museums and universities, well-paying and interesting jobs, and shopping and modern transportation systems. There is a point, though, where the advantages of urban density and high-rise living begin to diminish. At that point the idea of meta-cities and the creation of satellite cities that are linked to central cities by broadband communications and IT systems begin to make more sense. In the age of the revolution, telecommuting to work and tele-presence may breathe new life into the idea of meta-cities that can avoid the hazards, inconvenience and polluting aspects of rush hour travel to work on clogged highways and urban streets.

There are many advantages that can come with urban living in the future, but there are also significant problems and drawbacks as well. Cities, by acting locally, can help to reduce pollution, adopt a wide range of sustainable and green practices, including recycling more and more materials, reducing pollution and moving to what might be called an optimized meta-city form of urban living. The future of work and how super-automation transforms the nature of employment in the twenty-first century may well be the largest economic and political issue of the coming years and the Fourth Wave economy. New taxation policies related to the implementation of super-automation, AI technologies and the coming Fourth Wave economy are key issues that all industrialized countries need to face. The idea of a living wage for all people is increasingly being discussed in Europe and other parts of the world. In the megacities of the future, where machines perhaps will perform most jobs in the world – in farming and mining, in manufacturing, and in all forms of services – what will people do? This is perhaps the number one unanswered question about the cities of the future.

Chapter 5
Climate Change and Sustainable Development

Climate change is now affecting every country on every continent. It is disrupting national economies and affecting lives, costing people, communities and countries dearly today and even more tomorrow.

U. N. Sustainable Development Goal 13 on Climate Change

At the pace the world's population is growing, if left unchecked the world could reach 12 billion by 2100. To put this in perspective, in the year 1 AD (or CE, common era) the world stood at 200 million. By year 1000, 275 million. In 1800, 800 million; 1900, 1.6 billion. And by 2000, just over 6 billion. Can the world sustain such rapid growth.

Joseph N. Pelton in MegaCrunch: Ten Survival Strategies for the 21st Century

Introduction

The world's political leaders have placed their hopes to combat climate change in the Paris Accords that went into effect on November 4, 2016. This document, officially known as the U. N. Framework Convention on Climate Change (UNFCCC), did represent a historic achievement. The United Nations official account of what this convention is supposed to achieve reads as follows:

...strengthen the global response to the threat of climate change by keeping a global temperature rise this century well below 2 degrees Celsius above pre-industrial levels and to pursue efforts to limit the temperature increase even further to 1.5 degrees Celsius. Additionally, the agreement aims to strengthen the ability of countries to deal with the impacts of climate change. To reach these ambitious goals, appropriate financial flows, a new technology framework and an enhanced capacity building framework will be put in place, thus supporting action by developing countries and the most vulnerable countries, in line with their own national objectives. The Agreement also provides for enhanced transparency of action and support through a more robust transparency framework.[1]

---

[1] U. N. Framework Convention on Climate Change http://unfccc.int/paris_agreement/items/9485.php

© Springer Nature Switzerland AG 2019
J. N. Pelton, *Preparing for the Next Cyber Revolution*,
https://doi.org/10.1007/978-3-030-02137-5_5

In order to celebrate this historic occasion on November 4, 2016, the Eiffel Tower and the Arc de Triomphe were floodlit green. But the enthusiasm that accompanied the Paris Accord was brought crashing back to Earth when the climate change doubter, President Donald Trump, moved in 2017 to withdraw America from these accords and brought into question the viability of this treaty that every other country in the world had signed and many countries have now had their legislatures formally accede to it as well.

Population growth remains unabated. More people in more cities are serving to "fuel" (pun intended) the demand for more and more cars, houses and buildings that for the most part will still consume more carbon-based fuels and spit out more greenhouse gas pollutants. Many companies still do not have a clue as to how much greenhouse gas they emit, and most have not yet acted to create and implement viable plans to curb these emissions. Advances in photovoltaic solar cells, electric cars, and high-performance batteries fall far short of the Paris Accord objectives for a "new technology framework." In short, the world's 200-year increase in oil consumption and carbon-based fuels continues. If only one or two key countries pull out of the Paris Accord or significantly fail to meet its objectives, this key agreement could still fall apart.[2]

At the time that the accords went into effect, the story from a *New York Times* account expressed significant doubts as to likelihood that these new global agreements, signed by nearly 200 countries, would be able to achieve the stated goals. Indeed, at the time that the Paris Accords went into effect only 94 of the 195 countries that signed it had actually ratified it. Those 94 represented only two-thirds of the world's greenhouse gas emissions. (Note: As of 2018, the number had increased to virtually all of the 197 signatories. Perhaps most notably, the United States is the only country in the world that has withdrawn from this global agreement. With Syria's signature of this agreement, the United States stands alone in not being a party to it or pursuing its ratification). (Note: In addition Scott Pruitt, now the former administrator of the U. S. Environmental Administration, with Trump's backing, had zealously pursued rolling back all of the environmental regulations that were enacted in the Obama Administration).

The combined result is to say to the world that we are fine with the United States being the world's greatest polluter and have no concern for the sustainability of our planet, that we think climate change that gives rise to highly destructive storms or global warming are a hoax. These things, when taken as a whole, have not helped the standing of the United States in the world.

As noted in *The New York Times* story, significant likely gaps in the United Nations efforts in terms of the actual implementation were clearly identifiable.

The financial framework, namely a carbon price or tax that would force industries to pay for the pollution they spew, has barely started to emerge. And while tens of billions of dollars of green bonds have been issued to finance environmental projects, these are a pittance compared to the sums required to make a difference.[3]

---

[2] Keith Bradshernov, "The Paris Agreement on Climate Change Is Official. Now What?" *New York Times*, Nov. 3, 2016. https://www.nytimes.com/2016/11/04/business/energy-environment/paris-climate-change-agreement-official-now-what.html?_r=0

[3] Ibid.

Since the fall of 2016, the skepticism expressed by *The New York Times* article when the Paris Accords went into effect has proven to have been quite valid, as shown by the election of Donald Trump in the United States and the emergence of populist candidates in Europe that are skeptical of the need to reduce coal to generate electricity or the need to lessen oil consumption. Although the number of ratifications of the convention has now risen to about 80% of the countries that signed the Paris Accords, there are many signals to the effect that the UNFCCC is in trouble by failing to achieve its goals. And while billions of dollars for environmental project bonds have been raised, the scale of activity to truly thwart climate change and to meet global warming goals would, in fact, modestly require trillions of dollars.

The U. N. Millennium Goals and follow on Sustainable Development Goals (SDGs) are yet another clear-cut case in point. The U. N. General Assembly worked long and hard to create what were called the Millennium Goals for the turn of the century. These ambitious goals were clearly not achieved by the 2001 deadline. These goals have, in effect, now been recycled. This goal-setting process has conveniently morphed to become the Sustainable Development Goals for 2030. There are 17 of these goals and within them are 169 specific targets. The process of defining the goals within the United Nations was long and difficult, and the goals, as stated in the preamble to the official document, are noble and lofty. The objectives of the goals were for a global partnership:

> *All countries and all stakeholders, acting in collaborative partnership, will implement this plan.*
> *We are resolved to free the human race from the tyranny of poverty and want and to heal and secure our planet. We are determined to take the bold and transformative steps which are urgently needed to shift the world onto a sustainable and resilient path. As we embark on this collective journey, we pledge that no one will be left behind.*[4]

Over half of these 17 goals directly or indirectly address environmental goals for the air, the land, or the seas and thus in turn relate to coping with climate change.[5]

Actual progress to achieve these ambitious objectives remains far short of the stated aims. The sad truth is that realistically, when 2030 comes, the goals will remain unmet. Perhaps the action by President Trump to withdraw will receive the lion's share of the blame by environmentalists, but the problem of continued growth of population around the world, the continued growth of coal-fired electrical plants and an increasing number of gas-fueled cars signal a wider malaise in the fight to combat climate change (See Appendix 1 in this book).

This is because economic systems place a higher value on economic throughput, cheap energy and maximizing short-term profits. Corporate activity that dominates worldwide economic production and services does not recognize in its pricing systems or cost models the costs of coping with air, land, river and ocean pollution. In brief, economic goals and environmental/sustainability goals still remain non-aligned.

---

[4] U.N. Post 2015 Development Goals, August 12, 2015 https://www.un.org/pga/wp-content/uploads/sites/3/2015/08/120815_outcome-document-of-Summit-for-adoption-of-the-post-2015-development-agenda.pdf
[5] Ibid.

Private companies largely produce goods and services and governments are relegated to the role of cleaning up the consequences or creating the vital infrastructure or regulations to keep economic engines running. The issues that must be addressed range widely. These concerns can be pollution, over consumption of natural resources, transportation, national defense, disaster relief, weather forecasting, sanitation and water, labor issues, safety codes and inspections, health and education and more.

The process followed in most capitalist economies is that private enterprises "produce" and government "protects and regulates." As private enterprise becomes more and more predominant and provides the prime source of income and wealth, governments can and do become weaker and risk becoming more corrupt, more ineffectual at directing the activities of private enterprise, and less able to provide strategic leadership and cope with large-scale issues such as climate change.

The more technologically and scientifically complex, the longer term the issue or concern actually is, and the more investment that is required to address the problem, the less able governmental or regulatory processes are to address the problem effectively or in a timely way. Since technology and private innovation tends to be proactive and fast-paced in the world of technological innovation, while government tends to be reactive and slow in its reactions, the problem of disconnect between the world of corporate action and governmental response increases over time.

The classic example of this is the increasing pace of capitalistic enterprise innovation and automated throughput and the gap in time needed for governmental regulatory response. When one moves from national enterprise and government to global enterprise and global government, the issue of disconnect between industrial proactive programs and innovation and international regulatory response becomes even more difficult. The issue represented by ever-expanding and diversified industrial activity in a world populated by more and more people and the resulting climate change is a classic case in point. In terms of an economics problem, it is very much akin to the problem that all the ships at sea would like to have lighthouses to warn them of danger, but individual owners of the ships are unwilling to pay for the building of a particular lighthouse.

Initiatives for Coping with Climate Change and Implementing Sustainable Development

The U. N. Framework Convention on Climate Change (FCCC) and the U. N. Sustainable Development Goals for 2030 represent just the start of a process. These are merely goals to shoot for and very far from a panacea that can solve the gigantic challenge of achieving climate stabilizing and the halting of runaway global warming. The initiatives outlined below will need major progress in all five of these critical areas if Plan A is to save Earth as a livable planet that can sustain humanity for the longer term. We do not have centuries, but merely decades, to save Spaceship Earth.

Fig. 5.1 The U. N.'s seventeen sustainable development goals for 2030

Moderating Population Growth

The number one cause of climate change is the exponential growth of the human population, particularly since 1800. Figure 5.1 indicated the rapid growth of humans on Planet Earth in billions. The population was only 800 million as of 1800 but it is projected to grow to somewhere between 10 billion and 12 billion by 2100. In 1700 the world was perhaps 3% urban. Today it is 53% urban, and by 2100 it may well be between 75% and 80% urban[6] (See Fig. 5.2).

Today there are some 37 megacities with populations of over 10 million, but the number keeps increasing, and by the end of the twenty-first century the total number could be greatly increased[7] (See Fig. 5.3). Currently the rise in the number of people living in cities is far outstripping the total global growth of human population. In the next 40 years, global population growth will likely increase by around 1.5 billion people, but the increase in urban populations will likely exceed two billion.

This urbanization trend is key for several reasons. Cities are much more vulnerable to so-called "Black Swan" events, such as an asteroid strike, massive solar storm, or widespread failure of electrical power, communications or transport systems. But, in terms of climate change concerns, urban dwellers have higher levels of income, have more cars per capita, commute much further to work, and have higher rates of air conditioning. Particularly in developing economies, where there is often

---

[6] Joseph N. Pelton presentation, The Emerald Planet Television Program, January 8, 2012.

[7] World Urbanization Prospects, Department of Economic and Social Analysis (2014). https://esa.un.org/unpd/wup/Publications/Files/WUP2014-Highlights.pdf

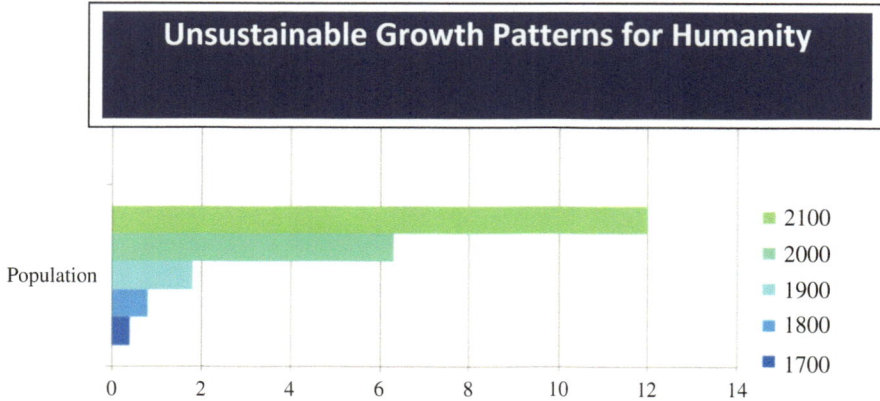

Fig. 5.2 Dramatic growth of human population worldwide (in billions) 1700–2100

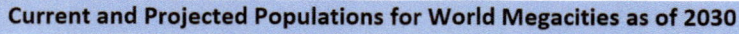

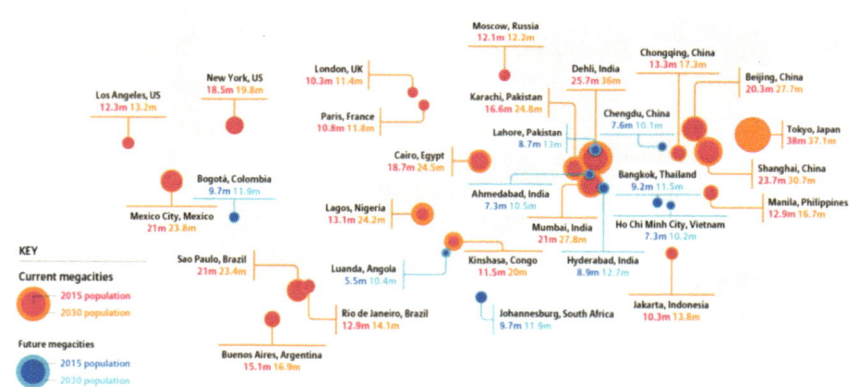

Fig. 5.3 The megacities of the world and their increasing populations. (Credit: U. N. World Urbanization Prospects, 2014)

an absence of electrification in rural areas and where industries are concentrated in cities, the power consumption and pollution rates tend to be concentrated in those cities. On the other hand, recycling programs, better insulation of buildings and homes, and other controls on pollution are often in effect. The full effects of urbanization versus rural life with regard to climate controls are in need of further study, but clearly the issues and the need for effective regulation vary greatly between town and country.

In some towns, and especially some smart cities with energy and air pollution plans, urban centers have developed effective strategies. Cities with a central regulatory authority over a specific footprint are more able to move toward a zero carbon footprint than in much more decentralized rural areas. If well-conceived energy, transportation, and environmental plans are devised and implemented in urban areas then they can address climate change issues. The Intelligent Community Forum (ICF), among others, is creating a global database of cities that have developed best practices in terms of telecommuting, use of renewable energy and energy efficient systems and pollution controls. But regardless of whether an area is urban or rural, or whether there are environmental controls, significant population increases trumps everything else.

Many industrialized countries have managed to achieve zero percent growth (ZPG) rates in recent years, but several Middle Eastern and African nations (such as Yemen, UAE, Burkina Faso, Niger, Nigeria, South Sudan, etc.), have growth rates above 2.5% per annum. India with 1.3% per annum growth rates will soon overtake China (i.e., growth rate of 0.43%) as the world's largest country in terms of population. Nigeria, with a growth rate of around 2.7%, could also overtake the United States as world's third largest nation in coming decades. The surging population in many developing nations represents the largest challenge to controlling climate change from a global perspective.[8]

The remarkable economic growth of China, which has allowed its per capita income to increase by an incredible 16 times, is due not only to rapid industrialization and economic expansion over the past five decades but also due to control of population growth.

The tough question is how to achieve a leveling off and perhaps ultimate reduction of the total number of humans on the planet? How does one slow down the growth of people that are driving more cars, consuming more oil, gas, uranium and electricity, living in houses and apartments, consuming natural resources, eating more food, and working on farms and mines and in plants and offices? Each human that is born in the world today releases many tons of greenhouse gases into the environment. A baby born in an economically developed country will release over 10 times more GHGs than a baby born in a developing economy. As we achieve the economic development aims in the U. N. Sustainable Development Goals to end poverty around the world, the problem of climate change becomes that much worse. The only longer-term solution is clearly some type of population control across the world.

Fortunately, the higher the rate of economic development, the lower the rate of population growth, and this is a good thing. But this slowing of population growth is insufficient. The answer must be found in additional ways. Possible answers include the following:

---

[8] World Population Prospects, 2015 https://esa.un.org/unpd/wpp/Download/Standard/Population/ also see. *World Fact Book*, May, 2017. https://www.cia.gov/library/publications/the-world-factbook/fields/2002.html

- Improved technology that makes birth control easier for both men and women.
- Change of religious prohibitions and practices that may have had purpose and utility centuries ago but are now no longer valid, and potentially destructive to the environment.
- Change to tax laws akin to those successfully used in Singapore that provides tax relief for one child but removes tax exemption with two children and imposes a tax penalty for more than two children.
- Tax and program incentives to villages in developing countries where population growth remains below target levels for zero population growth.
- Governmental or foundation support and incentives for people to not have children or waiting to have children until after 30 years in age, or for volunteering to undergo medical procedures to prevent the ability to have children or remove this capability for a fixed period of time.

Phase Out of Carbon-Based Energy Sources

The use of wood, coal, petroleum or other carbon-based fuels creates a number of problems. These issues include, among other factors: (i) the problem of air, land and water/ocean pollution that drives climate change; (ii) the fostering of illnesses and diseases of several types; (iii) the mining, drilling, fracking, and other means used to obtain these fuels, which represent hazardous employment; and (iv) the depletion of these natural resources for humanity's future use since, for the most part, they cannot be recycled.

Several fundamental propositions are quite clear. All energy consumed on Planet Earth essentially comes, whether directly or indirectly, from the Sun. Thus, as a logical proposition, we should set as a strategic goal for ultimate human energy use the objective of going to a 100% reliance on renewable and clean energy. This in time can be for all uses, including heating, cooling, transportation, networking, farming and industrial production. The goal is finding the right technology, one that is cost effective, portable and convenient to use. Since solar energy is available for free and in a virtually infinite amount, the issue is mostly envisioning and inventing the best technology to make this happen.

The proposition ultimately comes down to the fact that the use of carbon-based fuels is destructive of natural resources, polluting, a source of illness and hazardous work, and will be increasingly costly as these resources become harder to find and exploit. Ultimately this rather simple analysis reveals that a coal and gas economy represents a very poor way to obtain energy. Coal and gas interests spend billions of dollars to prevent the public from recognizing: that there is a clear cut need to phase out the oil, gas, and coal economy during the twenty-first century and replace it with some version of renewable solar economy. This is a plain as day. It is ironic that the economies of the Middle East that are extremely rich in both oil and solar energy resources have been among the first to recognize this logic and are seeking to lead this revolution forward.

The problem with this vital and essential transition in the world's carbon-based energy economy is the enormous legacy represented by jobs, industrial plant investments, automobiles, trucks, buses, trains, planes, ships, and devices that run on coal, gasoline, or oil-related products. Multi-billion dollar corporations with oil holdings all over the world supply fuels through worldwide distribution systems that support billions of different transport systems. Millions of people work to keep this vast carbon-based energy system running and billions need it to supply their cars, trucks, buses, trains and planes. This means that trillions of dollars will need to be invested over decades and decades to wean ourselves off this "dirty" and "dead end" energy system. Incentives – and some disincentives – will be necessary to make this switch. Closing down coal-based electrical generators is job one. Switching over to hydrogen-fueled or electric cars and vehicles is job two. A massive transition plan needs to be developed to figure out what to do with service stations, old coal-fired electrical generation plants, cars, trucks, trains and planes that run only on gasoline, diesel and/or jet fuel. This is a challenge both to national economies and resource planning and to global coordination.

A New Technology Framework

A new technology framework is a key step forward to achieve an effective and seamless transition to a new renewable energy economy. This transition is essential to creating a world that contains the worst ravages of climate change. There is a clear need for disruptive new technologies that allow the shift to a renewable energy economy over the next 30–40 years. It is hard to precisely identify these new disruptive technologies because most truly breakthrough advances are not an extrapolation from the past but rather true innovations that are a leap ahead and that are truly new and imaginative. Nevertheless some of the changes that might make a difference are proposed below for consideration (See Table 5.1).

The new technology framework as partially described in Table 5.1 has the potential to transform our world. Such initiatives could support new energy systems that could contain the ravages of climate change. Such new technological systems will be shaped by many factors but must also be curtailed by careful consideration of unintended longer-term environmental effects.

The point is that human technology has reached a new plateau of capability unparalleled in the history of humankind. The revolution in NewSpace technology or Space 2.0 is creating powerful new space systems, including new telecommunication satellite and remote-sensing constellations, allowing new on-orbit servicing and manufacturing and much lower cost and reusable launch systems. Tomorrow those same new capabilities in outer space might be used for even more daring and longer term goals, such as saving Earth from the ravages of climate change or creating a new atmosphere to change Mars from red to blue and green. But these are tools to be used with care and forethought. We have learned from the experiment called "Biosphere II" that humans have not yet been able to

Table 5.1 Examples of possible new technologies that could transform global energy systems, help contain climate change, and assist with planetary defense against cosmic hazards (Table developed and copyrighted by J. N. Pelton with all rights reserved)

Candidate programs for a new technology framework for climate change alleviation
Developing advanced, lower cost and more efficient solar energy electrical energy conversion systems such as those using quantum dot technology
Development of new clean powering systems for vehicles that are hydrogen-fueled, using advanced and lower cost fuel cells or compressed air
Creating a solar power satellite system to beam down clean energy
Building a solar shield to modulate the amount of solar energy reaching Earth and ward off solar storms that threaten modern electronic infrastructure
Creating a new ozone layer for Earth
Moving Earth's orbit outward to reach further from Sun
Painting of clouds white
Simulating volcanic eruptions. (Sulfuric acid clouds reflect sunlight and thus cool the planet)
Freezing the peat fields of Siberia and capping Arctic methane plumes
Chemical treatments to change the icecap albedo
Increasing iron-enriched ocean bio-productivity
Installing global "heat pipes" to eject energy into space
Performing genetic altering of cattle to reduce methane emissions
Improving birth control technologies and giving tax incentives for human birth control
Re-engineering of vast mining operations in Siberia to cap methane emissions
Creation of U. N. Scientific Council on New Technologies to control climate change

create a sustainable self-contained environment, and designing space tools to save Earth could turn into a dangerous experiment indeed (Fig. 5.4).[9]

Perhaps the most significant one of all is the advent of the so-called Singularity. This is the name that was devised by AI guru Ray Kurzweil, who developed the software used by Siri of smartphone fame. This term seeks to highlight the dramatic changes that will occur to human society when we develop "artificial brains" akin to the reasoning power of humans and they become widely available to provide services and help devise new technologies for all modern societies.

We are beginning to see precursors of this revolution in the use of IBM's Watson technology to assist in everything from filing income tax returns to diagnosis of medical ailments. Already Watson is smarter than most CPAs and more accurate than most doctors in determining rare diseases. Already researchers have developed the equivalent of the brain of a rat, and it is thought that an artificial human brain with algorithms equivalent to the reasoning power of a very credible scientist might come within the next two decades or so. The technology that teams of AI brains might devise in the not too distant future might be impressive indeed.[10]

The advent of artificial, intelligent, logical brains will transform the innovative process. Creating the logical processing machinery to make this happen will be very

[9] J. L. Green, J. Hollingsworth, D., Brain, V. Airapetian, A. Glocer, A. Pulkkinen, C. Dong and R. Bamford, "A Future Mars Environment for Science and Exploration", Planetary Science Vision 2050 Workshop 2017 (LPI Contrib. No. 1989) 2017.

[10] Ray Kurzweil, *How to Create a Mind: The Secret of Human Thought Revealed* (2012). Penguin Press, New York.

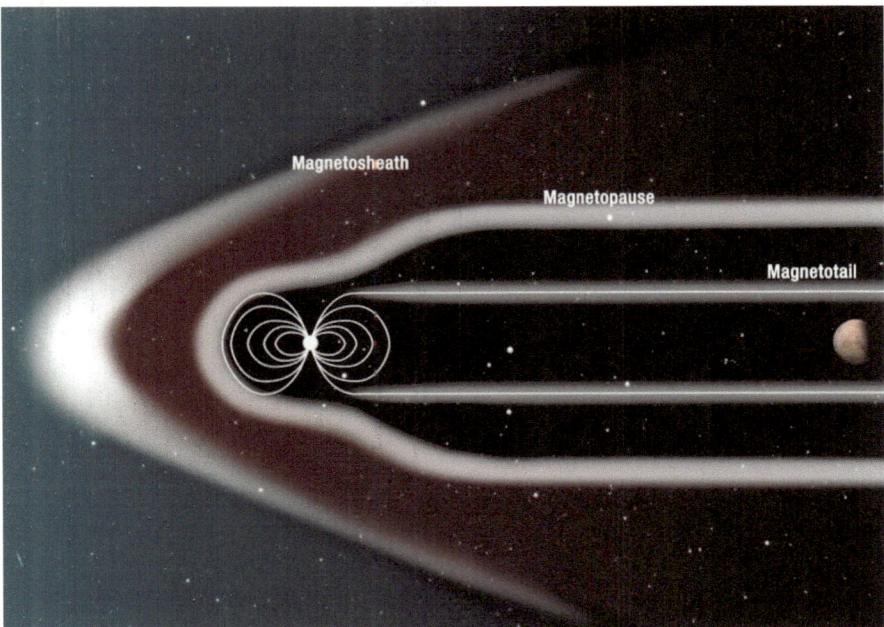

Fig. 5.4 NASA study concept for a magnetic shield for mars to allow an atmosphere to form. (Graphic courtesy of NASA and Dr. James Green)

difficult indeed. Creating the intuitive algorithms that allows for artificially intelligent innovative thought will be even more difficult. Perhaps the greatest challenge of all will be convincing slower-witted human political leaders the wisdom of using creative new technology to solve very difficult problems such as climate change, global warming and other key challenges such as offering universal health care and education at greatly reduced cost.

Strategies for Developed and Developing Economies

The need to cope with climate change will become increasingly clear as violent storms, tornados, hurricanes, typhoons and other natural disasters become more prevalent, water shortages become more severe, and younger people are educated about the nature of climate change and the dangers it spawns. The increasing release of methane from melting peat bogs in Siberia and the Arctic regions and more and more greenhouse gases released from along ocean shore regions will also make the effect ever more noticeable.

The strategies to address climate change concerns and the effects related to global warming, such as more violent storms, changing weather patterns, and more droughts and altered rain patterns, will require responses around the world. These strategies in terms of legal reforms, policy changes, regulatory reforms, and new taxation and economic programs will vary around the world. They will be significantly different for countries with developed, industrialized economies and those with developing economies.

There are two very significant problems to be addressed in developed economies. First, most people in these societies are so accustomed to a standard of living related to food, housing, heating and air conditioning, and travel by personal cars and aircraft that they are reluctant to give up these luxuries of life in order to accommodate what appears to be needed to combat climate change. Second, the various business enterprises that are involved with carbon-based fuels, conventional energy, transportation, housing and buildings, and industry are heavily invested in infrastructure and energy systems that are driving climate change.

Many of those people who understand the problems associated with climate change are nevertheless generally unwilling to accept the degree of change that is necessary. They are unhappy to pay the cost in terms of taxes and altered conveniences that may be required. They are happy with a diet based on energy-inefficient heat, air-conditioned homes and so on. They are resistant to changing the conventional way of doing things. They have yet to be truly convinced that all communities and all people must change. These people still need to be totally convinced that we must urgently invest in new renewable energy systems and that new, environmentally friendly and clean technology must substituted for the old.

Secondly there are the vested economic interests. Many investors and business interests are likely going to resist the legal, technological and regulatory changes that are going to be needed. These economic opponents include owners of oil and gas fields and coal mines, refineries, of car and truck manufacturing plants, of service stations, and even of rental housing, apartment buildings, office buildings, and shopping malls. These business leaders and investors have yet to figure out how to counter climate change without suffering huge economic losses. The idea that there is going to be a 40–60-year transition period until clean renewable energy systems are available seems to be fine for them, but – and this is a very key 'but' indeed – only if they personally are not asked to endure significant financial losses during that time. And many workers also worry about whether the battle against global warming or climate change might lead to the loss of their job in a year or two?

Conclusions

The issues of climate change and environmental pollution are worldwide ones that are increasingly a threat to the world economy. If the reforms we have discussed are not truly successful by the end of period that we have deemed the revolution, then the issue does not become an economic one but an issue of survival. Today most people see the issue as economic and environmental. By the year 2100 we will really know if we can save our planet from becoming a runaway fireball or we must depend on radical ideas such as creating a Sun shield to ward off solar radiation or painting Earth's atmosphere white to reflect the microwave-baking radiation of the Sun.

In developing societies there will be the need to affect change to combat climate change and environmental pollution through two main initiatives. The first are programs aimed at reducing population growth and relieving massive overcrowding. Further there need to be new incentives and penalties (i.e., carrots and sticks) that perhaps work best at the level of villages and towns rather than via individual measures such as tax policies.

In economically developed areas the dynamics are greatly different. There are at least two types of problems. On one hand, there is the need to change and make some sacrifices of convenience by those who have been accustomed to not having to sacrifice. On the other hand there is the question of significant economic losses by embedded business interests who perceive the threat of major financial losses due to climate change reform. Millions of people are very reluctant to give up their meat-heavy diets, their large and energy-inefficient houses, their air conditioning and their gas-fueled cars. Further, wealthy and powerful business interests are inclined to lobby to keep and operate their coal mines, their oil refineries and their inefficient hotels, apartment buildings, shopping malls and vehicle production plants.

These things can and will change in time. Reforms will be made at the local, state, regional, national, and even global level. The question is can they change quickly enough? Can the political and business interests recognize that change is truly urgent? Do they understand the devastating scope of the losses that are coming due to global warming, sea level rise, powerful storms fed by the energy put into the oceans and the atmosphere, etc.? Will they realize that paying for conversion now is going to save huge costs later as the icecaps melt and Siberian peat fields release massive amounts of methane that rise into the atmosphere and become incredibly difficult to re-extract. Once icecaps melt into the sea, the seawater is not able to easily refreeze back again. The difficulty of reversing climate change mega-shifts that come from released methane and melted ice is hard to grasp. Thus, climate change issues remain perhaps the most confounding hyper-object problems for most people around the world. Closely tied to the issue of climate change and the difficulty of pollution controls, however, is the issue of exponential growth of human population and the need for limited population expansion to provide the opportunity to control climate change, global warming and the excessive extraction and use of resources.

Chapter 6
Life in a Cybernetic World

Any sufficiently advanced technology is indistinguishable from magic.

Arthur C. Clarke's Third Law

Welcome to the Cybernetic Revolution

The world we live in is a cauldron of change. It used to be that change in basic societal conditions and indeed any major shifts in key patterns of human life took thousands of years to occur. Then it was hundreds of year, then decades, followed by change that only took years to occur during the twentieth century. Now, in the cybernetic world change seems almost instantaneous. In this age of artificial intelligence, the Internet of Everything and social media. Societal shifts occur over months, weeks, days and sometimes hours. This destabilizes family life, education, health care and the socialization of youth as well as the generational gap between the young and the old. It can affect the nature and type of criminal behavior that can and does occur. And, as we have seen in the case of electronic meddling in elections, it has had an impact on political processes in the United States, France, Germany, Hungary, and Brazil – to name just some of the instances where "weaponized fake news" and bot-based false identities have shaped public opinion, introduced racist hate messaging and distorted election processes. Another significant consequence is the loss of personal privacy.

This pattern of ever-accelerating rates of rapid social change was labeled by the author almost two decades ago with the term "societal jerk." It was in the context of discussing how information technology and rapid social change were transforming society. The use of this term first appeared in my book *eSphere: The Rise of the World Wide Mind* (2001). In the world of physics, "jerk" is experienced not when things are accelerating but when the rate of acceleration is itself increasing. In the world of physics "jerk" events that are measured in fourth order exponentials occur only very briefly. Jerk is something typically measured in nanoseconds or at most milliseconds, such as in bombs exploding, rockets taking off, or when your head

© Springer Nature Switzerland AG 2019
J. N. Pelton, *Preparing for the Next Cyber Revolution*,
https://doi.org/10.1007/978-3-030-02137-5_6

snaps back when you stomp on the accelerator of a hot rod. In the post-information age of the emerging "World Wide Mind" it has been happening for quite a few years. Jerk is by definition disruptive. Sustained societal jerk now engulfing our global information systems has zoomed from being measured not just in terabytes but now in yottabytes – one septillion bytes of information, or 10^{24} bytes.

One reality of this transformation of globally stored massive databases is that they are too large for humans to relate to anymore. Only supercomputers can comprehend or reasonably access databases measured in yottabytes.

Today, modern society is being "jerked" into new patterns of behavior by personal communications and social media, and soon by 'smart' things created by the Internet of Everything, which will be talking to people and increasingly to each other.[1] The Internet of Things will become the Internet of Everything, and human social environments will involve people, animals and a rich, interactive world where we communicate with "smart things" almost as with other humans or pets. We will be constantly talking to and responding to our automobiles, trucks, buses, trains and airplanes.

Our lives will be shared with appliances that will be absolutely chatty. Refrigerators will remind us about food we need to reorder and washing machines will tell us when to use more detergents or bleach. Cars will tell us when we need to take them in for service, replace our tires, or choose a better route to drive to avoid traffic tie-ups. In fact, our self-driving cars will automatically choose where and how we drive across town or across country. In this new life, we will likely converse more with television sets, toasters, microwave ovens, security systems and baby monitors than we do with friends and family. Life within the world of the Internet of Everything, and a universe driven by social networking with smart cars and appliances, will be disorienting and perhaps even at times distressing. This plethora of information derived, stored and organized by the Internet of Everything, will also add to the loss of personal privacy.

And chatting with cars and appliances will be only a small part of the change. Education and health care will be a significant part of the change in this new world that is coming at us fast.

A Re-imagined World

Most aspects of education can be reduced to efficient media labs that allow students direct interaction with machines that teach them everything from the alphabet to basic mathematics to reading skills and more and to do so much more efficiently than classroom teachers and at a small fraction of the costs. Video labs that allow students to learn at their own rate and complete online homework assignments and take competency tests at accelerated rates can revolutionize education.

---

[1] Joseph N. Pelton, *E-Sphere: The Rise of the World-Wide Mind* (2001), Bridgeport, Conn.: Quorum Press.

Friend, tutor and wise advisor Arthur C. Clarke made a keynote speech at the UNESCO inaugural opening event for the International Programme for the Development of Communications (IPDC) at its Paris headquarters in the 1980s. At the time, he prophesized a future in which electronic encyclopedias could be connected and updated around world to bring the latest information to students everywhere. He noted even then that the type of technology he was envisioning had more to do the with the electronic toymaker "Mattel" than with IBM. At the time he was confronted with a somewhat hostile press that asked: "Are you saying that teachers should be replaced by machines?" Clarke explained that no, there would still be a need for teachers, but that his low cost and very knowledgeable computerized sources of information could greatly expand the scope and speed of learning. He also added: "Come to think of it, all the teachers that can be easily replaced by a machine probably should be."

And if smart machines and robotic educators and interactive video labs can revolutionize the speed, scope and range of education and training, then healthcare will also be revolutionized. Fitbit watches and I-Phone health monitors are just the first wave of the coming change to healthcare as we know it today.

Currently, under most medical systems around the world, people go in for a checkup just once every 2 years, or perhaps every year as they get older. Major tests such as colonoscopies are conducted every 5–6 years. But there is evolving smart medical technology that could allow the constant checkup on a person's health 7 days a week, 24 h a day.

For many years, smart satellites have been providing advanced communications or remote sensing in the skies using something called autonomous operation. These satellites sent telemetry streams to computers that reported on temperature levels, electrical current flows and data that indicated healthy operations. If any anomaly occurred with the dozens of key elements of their operation, then corrective action could be taken either by computer programs or by ground controllers who could use override to restore the satellite to optimum health.

With the knowledge now developed by heuristically smart computer systems such as the IBM Watson, trained in advanced medical diagnosis, it would be entirely possible to monitor human health on a continuous basis and apply the appropriate medical treatment without a doctor or nurse being a part of the process, except in very rare circumstances. Most ailments such as flus, colds, skin conditions, breaks or strains, etc., can be sorted into perhaps a hundred or so categories with specific responses to which dispensing robotic systems could easily respond with a treatment. For that matter medical treatment could be issued via universal service code to pharmacological machines. These robotic devices could dispense medicines that are charged to credit cards so that the great majority of pharmacists and pharmacy systems could be dispensed with – if one can permit the unfortunate pun.

It might take a decade or two to develop, trial test and fully implement a system that includes automated health monitors that would be attached to all of a citizenry's arms and perhaps wired to the heart, brain, or other glandular systems. Such systems can be designed for aircraft and artificial satellites, and ultimately they could be designed and implemented seamlessly to all humans to detect any sign of illness, injury, ailment or aging and to prescribe with great precision the appropriate response and medical treatment. This is not to say that such an automated system will not raise significant issues and treatment concerns.

Such an automated and robotically enhanced capability when applied to education and healthcare could greatly increase the effectiveness and cost efficiency of these activities that claim 25% of total economic costs incurred by most developed countries in the world today. A great deal of the savings would be, of course, in the elimination of labor costs. Such a system might ultimately reduce the number of teachers, professors, doctors, nurse practitioners and other professionals and assistants in the affected fields. Some of the reductions over time might be on the order of ten to one or even twenty to one in comparison to today's employment levels. Even those that remained in the field of education and healthcare would be reduced more and more to the level of technician or human-relations expert or advertising adman with a resulting significant decline in income.

The Automated World and Civil Liberties

There are lots of things that people do that an automated medical monitoring system would find contra-indicated. These practices such as smoking, using drugs, engaging in unprotected sex, not obtaining sufficient sleep, even engaging in extreme sports, would come up as unhealthy practices. It is not at all clear how such "infractions" would be addressed in an automated medical monitoring system. Would there be increased medical payments incurred by those that engage in unhealthy practices? Would penalties progressively go up as infractions occurred? Would such health infractions be reported via an automated health monitoring system that would require first counseling, then fines, and perhaps even sterner actions? Would unhealthy personal practices even possibly include public censure or refusal to provide medical care?

And this just the start of liberty concerns that arise might from an automated medical monitoring and health services system of the future. There could be other forms of societally imposed sanctions that could come from automated healthcare monitoring systems that could seem to be very Big Brotherish indeed. Talking cars and toasters might require a bit of adjusting to, but what will you think when your automated healthcare system tells you your heart or lungs are "out of specified limits" and you must immediately rest, take a nap, or report to a hospital.

It would be possible that one's personal health monitoring system could even be used against you in a court of law. A complete record of one's respiration, heartbeat and glandular activities – perhaps linked to GPS location records – could be used to convict a rapist, murderer, or robber. It might even be used in divorce cases to prove infidelity:

> *Your honor, I submit as exhibit 12 Mr. X's and Mrs. Y's medical and GPS records. These records clearly show that he was at the abode of Mrs. Y. During the date in question between 21:05 pm and 21:15pm Mr. X's and Mrs. Y's respiration, heart rate, and glandular secretions are completely consistent with engaging in sexual intercourse followed by*

ejaculation by Mr. X and then a return to normal bodily functioning. We can submit the records of laboratory treatment subjects who are engaging in sexual intercourse covering at least ten different couples – or more if required. These medical records show complete consistency in the monitored bodily functions for both Mr. X and Mrs. Y – and with a 98% consistency.

Such automated medical recording systems and who might have access to them and for what purposes would undoubtedly raise perhaps hundreds of new issues related to civil rights and civil and criminal law. The American Civil Liberties Union, among many others, might feel the need to protest and object to a "Brave New World" in which automated medical care systems replaced medical care practices as we know them today. What we do know is that these systems could be used today, especially in rural and isolated areas where doctors and hospitals are in short supply.

On one hand we know that today's medical practices are truly antiquated. They are costly and not nearly as efficient as they could be if the latest technologies and healthcare monitoring systems were systematically used. But such automated and universal health monitoring systems if uniformly required would come at a high price in terms of autonomy, liberty and civil rights.

In short, the true expertise, knowledge and even research competencies would reside much less with humans and much more with heuristically sophisticated machines with high levels of artificial intelligence and access to yoddabytes of information to support their educational, healthcare or medical pursuits. Today, IBM's Watson, which is assimilating terabytes of medical pathology, diagnostic technique and medical research and test information, seems like a quaint anomaly, but tomorrow the spread of thousands of IBM Watsons to hospitals and medical centers will change the world as we know it today.

And as we move forward to use more and more automation, more heuristic algorithms to make diagnoses and more artificially intelligent monitoring systems, there will not only be issues related to civil liberties but also of cybersecurity. In the future not only secure databases can be hacked, but medical records, automated medication dosages and even implanted medical devices could be illegally accessed to do people harm or even kill patients via cyberattacks, either as a criminal act or as terrorist attack.

Are We Destined to Live the World of Kurt Vonnegut, Jr.'s *Player Piano?*

One of the most famous lines that derive from the cerebral mind of Arthur C. Clarke is that "Almost anything we can conceive of will likely eventually become reality."[2] Certainly there are a number of dystopian science fiction novels that portray a future where humans are subjugated to a life without real meaning,

---

[2] Joseph N. Pelton, *The Oracle of Colombo: How Arthur C. Clarke Invented the Future* (2015) Emerald Planet, Washington, DC.

where there is a lack of creative and interesting work, and there is an oppressive political existence. The Vonnegut novel *Player Piano* portrays a future in which the great masses of human population have become a proletariat of consumers of low-quality products. The great majority of people no longer have much of a real mission in life other than to procreate and to buy shoddy products turned out by robotic factories that are supported by a small class of scientists and engineers. The existence for protagonist Paul Proteus, manager of a mammoth automated plant, is not all that much better that the proletariat. His life is not so much oppressive as dreary and without any real purpose.[3] Nor is *Player Piano* unique in the world of dystopian science fiction in imagining a future where technology not only does not liberate humankind but rather serves to take the joy out of daily life for most humans.

And one does not have to turn to sci-fi dystopian literature to find significant concern about how broadband networking, the Internet of Everything, artificial intelligence and heuristically enabled robotics might undercut our pursuit of happiness. The 1971 book by William Kuhns entitled *The Post-Industrial Prophets: Interpretations of Technology* presents a series of critiques against automation and the regimentation that can come from technological innovation without a moral compass.[4] This interesting book, that seeks to interpret technology, presents the philosophic insights of such thinkers as Jaques Ellul, Siegfried Gideon, Lewis Mumford, and Harold Adams Innis. These skeptics present a range of arguments showing how advanced technology and automation, particularly when driven by strictly free-market capitalism and profit motivation, can lead to a lessening of quality of life and regimentation and restrictions on freedom, liberty, and personal innovation.

Sigfried Gideon, for instance, explains the downsides of automation of bread production. In this case continuous belts push bread dough through vast long ovens to produce a baked product as it emerges a minute or so later at the other end. The profit-driven automation in this case results in ever cheaper but lower quality and harder to digest loaves of bread. This example, he suggests, is just one of many paradigms regarding how profit-driven and "efficiency-of-rapid-production-at-all-costs" will always degrade the mass market and in the longer term degrade the quality of life for humanity.

The uses of technology in society today are driven by a limited number of objectives. These uses essentially fall into two categories: (i) market expansion or (ii) productivity gains to support market expansion. Thus the modern uses of technologies include envisioning and creating more effective weapons and armament systems, producing an expanded range of consumer products or social services (i.e., primarily government services, education, or healthcare) that can be constantly improved and upgraded to fuel growth of global consumer products and/or services, or expanded demand for tax-supported governmental services. In support of the ability to create these expanded markets and also support demand created by

---

[3] Kurt Vonnegut, Jr. *Player Piano (1974)* Dial Press trade Paperbacks, NY.
[4] William Kuhns, *The Post-Industrial Prophets: Interpretations of Technology* (1971) Harper Colophon Books, NY.

expanded populations there is also the drive to create enhanced technological services or products that can increase worker productivity either through automation, improved supply chains, or improved sales and distribution efficiencies.

It is possible, of course, to envision other uses and purposes for technological innovation. These would include such goals as improved knowledge of the universe and its composition and functioning; to create a better quality of life for all humanity; to reduce pollution and create a more long-term sustainable world; to eliminate wars, ethnic strife, racism and religious intolerance; to reduce economic hardship and disparity; to increase longevity and decrease all forms of suffering; or to allow governments to be more democratic and responsive to the needs and aspirations of their citizenry.

Essentially the entire world is driven by economics and market expansion. The prime use of technology in all modern economies – whether communist, free-market capitalist, or even totalitarian – is to serve economic goals, continuous market growth, and/or augmented military capability. The prime question that thus underlies the future world that the Cyber Revolution will bring to us is, what is going to be the outcome of humans and smart robots working together? Can humans and smart machines work together to create a better or a more repressive world? (See Fig. 6.1)

What Is the Real Threat of Smart Machines That Think Like Humans?

This is actually one of the fundamental questions facing human society today regardless of the fact that few people would actually deem this to be the case. And today the answers that come back from so-called experts are contradictory and largely confusing. Some, such as Elon Musk, have suggested that super smart but malevolent smart machines might take over and rule humans. This is a frequent theme seen in movies such as *I, Robot* or Skynet in the *Terminator* movies.

Fig. 6.1 The smart robot: a helping hand? (Graphic courtesy of Completewellbeing.com)

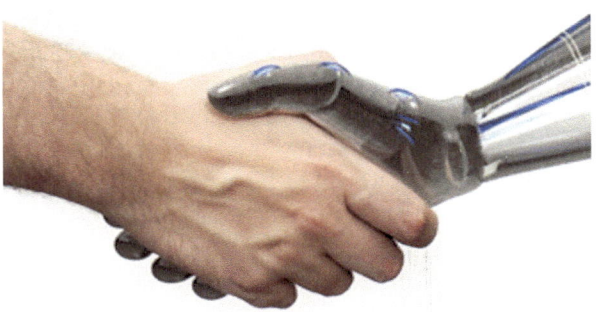

Others suggest that super-automation or the Singularity as envisioned by Ray Kurzweil could not only redefine work as we know it but lead to massive unemployment. Some advocates of the massive use of robotics and artificial intelligence say that automation and artificial intelligence just make workers more efficient, and that this technology will, in fact, create more jobs.

In the world of manufacturing, mining and farming we have seen job after job disappear and are increasingly finding that the new jobs that are being created on any large scale are those such as workers in fast food restaurants, hotel clerks and maids, teacher assistants and retail checkout clerks. The majority of new jobs tend to be low-paying minimum wage menial-labor service jobs that do not represent a career and are seldom rewarding or challenging on a longer-term basis. Today automation in farming, mining and manufacturing has largely run its course, and now it is smart machines that are replacing service jobs. Smart heuristic programs are beginning to replace appraisers, pharmacists, book designers – any job that follows a basic routine. Tomorrow, as noted above, automation will be taking on education, healthcare, automotive design and a vast array of service jobs.

Kai-Fu Lee, president of the Artificial Intelligence Institute, has made the following "reassuring" claim about the pathway forward in artificial intelligence: "At the moment, there is no known path from our best AI tools (like the Google computer program that recently beat the world's best player in the game of *Go*) to "general" AI – i.e., self-aware computer programs that can engage in common-sense reasoning, attain knowledge in multiple domains, feel, express and understand emotions and so on."

But even Dr. Lee, who provides this reassurance, notes that even today's smart AI tools are enough to change our world, redefine the meaning of work, and redistribute wealth.[5]

Today, the stereotype is that machines are cold digital calculators while humans are empathetic. Yet the odds are that if we can build machines that are someday soon as smart as humans, we can probably make them empathetic and perhaps more sympathetic than most people as well. It is interesting to speculate about the future of machines with intelligence, reasoning and judgment skills. Will such smart algorithms as they evolve within intelligent machines also be likely to develop an ego as well?

There are experts that would disagree with Lee about the near-term pathway forward. Professor Henry Markham in France, along with a very competent team, is working hard on creating a digital neocortex that they hope to be able to support 10^{17} bytes of memory and carry out 10^{18} digital logic flops per second by 2023. This would represent the full equivalence of a human brain. And some believe that the heuristic algorithms that work in tandem with this amazing artificial brain will be just as powerful as human logical processes. As Ray Kurzweil contends with regard to Pattern Recognition Theory of Mind (PRTM): "The human mind

---

[5] Kai-Fu Lee, "The Real Threat of Artificial Intelligence," *New York Times,* June 25, 2017. pp. 3–4.

seems limited to coping with four hierarchical lists of pattern recognition at a time. An 'artificial neocortex' would not need to be so limited."[6]

Are Universal Basic Income (U.B.I.) Payments Coming?

The megatrends that are implicit in the world of the Cyber Revolution are far from clearly understood. Most people do not understand what the onset of the Singularity will mean. They get it when it is suggested that smart robots could take over jobs that involve physical labor. Most people have seen robots at work on an assembly line. But few today really get what it means for white collar jobs held by college graduates to have thinking machines able to perform 24 h a day without pay, pensions, sick or maternity leave, vacation or occasional strikes for higher benefits or a demand for a corner office.

Today, in developed economies, 80% or more of all employees have service jobs that require a certain level of education, training, judgment, and perhaps management skills. What is it that people do and what kind of salary can they demand if smart robotic systems possess the intelligence and smart algorithms to do their jobs? When machines have the educational background, training and judgment and reasoning skills that allow them to be accountants, appraisers, educators, healthcare providers, truckers, rail road employees, sale representatives and more, what do people do and how do they earn a living?

This is a problem that is beginning to roll out almost in slow motion today, but the scope and scale of what is to come is largely dismissed by politicians and business leaders who have the most to gain by ignoring an unfolding reality. There are some with the insight, such as Microsoft founder and computer wizard Bill Gates. Gates has noted that workers are taxed for doing their work, and robots should be essentially taxed the same way. He has suggested that the tax could go to subsidize jobs that require human empathy, such as care for the elderly, healthcare and education that is underfunded.[7]

In 2017 the European Parliament considered and then rejected the idea of an EU-wide robot tax as a means of coping with the rising use of robots. There are many economic and practical reasons why this would be difficult to implement, starting with what exactly is a robot. The main point is that the key issue for the future is really not robots in factories taking jobs but smart algorithms that can and are taking over white collar jobs that have nothing to do with factories. What the

---

[6] Ray Kurzweil, How to Create a Mind: The Secret of Human Thought Revealed, (2012) Viking Press, NY, pp. 124–132.

[7] Daily Brief, "The robot that takes your job should pay taxes, says Bill Gates" Quartz, https://qz.com/911968/bill-gates-the-robot-that-takes-your-job-should-pay-taxes/ (Last accessed July 7, 2018).

European Parliament did do was to enact the start of "an ethical framework for their development and deployment and the establishment of liability for the actions of robots including self-driving cars."[8]

What is beginning to be discussed and even tested in some places is a policy for the future called a Universal Basic Income (U.B.I.). This is a payment that would go to all people and would replace all other social benefit programs such as food stamps, Social Security retirement benefits, etc. There has been one calculation of what this would cost in the United States on an annual basis if everyone were to receive $1000 a month plus catastrophic healthcare coverage. The total of such a direct payment system would come to the tidy sum of $3.9 trillion. Of course this would eliminate over half the cost of the federal budget associated with various entitlement programs.[9]

There are proposals by advocates such as Annie Lowry, author of *Give People Money: How a Universal Basic Income Would End Poverty, Revolutionize Work, and Remake the World,* who has suggested that a taxation scheme that included new levies on inheritances, a carbon tax, and new pollution taxes could make such a system work. Andy Stern, the former president of the Service Employees International Union, also has advocated such a system as well as a standardized 15-h work week as an answer to the rise of robots. Today there are pilot trials of Universal Basic Income payouts in Finland, Canada, the Netherlands, Scotland, and Iran and several local trials in northern California and Kenya, where success is being claimed.[10]

Although U.B.I. may not be the answer, some adjustments are going to be needed, and U.B.I. gives insight into how engaged journalists, union leaders, and thoughtful billionaires such as Bill Gates, Elon Musk and Sir Richard Branson are exploring the future of work and compensation in the brave new world that will come with the Cyber Revolution that is unfolding as the twenty-first century progresses.

Can the U. N. Sustainable Development Goals Change the World?

Within the United Nations there have been initiatives to change the dynamics of a world driven by industrial expansion and ever greater consumption and accumulation of wealth to a new type of world much more driven by eleemosynary – or what some would call more "noble" global goals. These U. N. General Assembly agreed noble objectives were first known as the Millennium Development Goals and more recently, after the first 15 years of the millennium had come and gone, were

---

[8] Reuters staff, 'European parliament calls for robot law', Reuters.com, February 16, 2017. https://www.reuters.com/article/us-europe-robots-lawmaking-idUSKBN15V2KM

[9] Nathan Heller, "Take the Money and Run" *The New Yorker,* July 9 & 16, 2018, pp. 65–69.

[10] Ibid.

renamed the Sustainable Development Goals. These U. N. goals have placed new emphasis on environmental sustainability, bridging the economic and digital divide, and employing technology to create a more equitable, fair and free world for all humanity (See Appendix 1 in this book for the more detailed description of these goals).

The problem is that these goals (see Table 6.1) are essentially aspirational rather than tangible and realistic goals that could truly be implemented by 2030. In fact, there are no implementation milestones or enforceable deadlines. The first of the U.N.'s attempts in this direction were the Millennium Development Goals that were to have been achieved between 2000 and 2015. This period has now come and gone, without the sincere 'buy in' by large business leaders, capitalists, entrepreneurs, and the political leadership of the world. Without the commitment of trillions of dollars of financial resources, specific targets and deadlines, and an almost revolutionary commitment by political and economic leaders, it seems likely that the Sustainable Development Goals for 2030 are likely to follow the same pattern of visionary goals followed by disappointment.

Table 6.1 The United Nations 17 sustainable development goals for 2030

The United Nations' seventeen sustainable development goals	
Goal 1	End poverty in all its forms everywhere
Goal 2	End hunger, achieve food security and improved nutrition, and promote sustainable agriculture
Goal 3	Ensure healthy lives and promote well-being for all at all ages
Goal 4	Ensure inclusive and equitable quality education and promote lifelong learning opportunities for all
Goal 5	Achieve gender equality and empower all women and girls
Goal 6	Ensure availability and sustainable management of water and sanitation for all
Goal 7	Ensure access to affordable, reliable, sustainable, and modern energy for all
Goal 8	Promote sustained, inclusive, and sustainable economic growth, full and productive employment and decent work for all
Goal 9	Build resilient infrastructure, promote inclusive and sustainable industrialization, and foster innovation
Goal 10	Reduce inequality within and among countries
Goal 11	Make cities and human settlements inclusive, safe, resilient, and sustainable
Goal 12	Practice responsible consumption and production
Goal 13	Take urgent action to combat climate change and its impacts
Goal 14	Conserve and sustainably use the oceans, seas, and marine resources for sustainable development
Goal 15	Protect, restore, and promote sustainable use of terrestrial ecosystems, sustainably manage forests, combat desertification, and halt and reverse land degradation and halt biodiversity loss
Goal 16	Promote peaceful and inclusive societies for sustainable development, provide access to justice for all, and build effective, accountable and, inclusive institutions at all levels
Goal 17	Strengthen the means of implementation and revitalize the global partnership for sustainable development

Attempts to redeploy technology to achieve environmental reforms may have been agreed on within the politics of the United Nations but not by the capitalists of the world that command most of the world's wealth. There are apparently those who still believe that they can take their money with them into death and that "too much money still ain't enough."

There is today a dichotomy of those who believe in creating a sustainable world versus those that believe this is not at all necessary and that environmental concerns are all overblown. In part, this lack of concern comes from Fundamentalist beliefs that a Supreme Being is watching over Earth and humanity and would never let the human race and a habitable world perish.

Another view is that is perhaps more logical would be that intelligent life on Earth is an experiment. If we cannot find a pathway to a sustainable existence and to create a circular economy based on recycling of our resources and not overpopulating our planet, then our species deserves to die out. Even if there is a Supreme Being who created the world, then it may be that humanity itself must take responsibility for sustaining that which has been created.

There are economic and political leaders – those that control most of the world's wealth – who seem to believe that there can be continued exponential growth based on capitalist expansion, an ever-growing global population and even larger levels of consumption of resources. They have simply not bought into the idea that the top goal in society should be to preserve Earth as a livable place to sustain the human race and the richness of diverse biological and natural resources. Fortunately there seems to be a growing recognition that runaway growth is not a sustainable view of the future. The more humans become big eaters, the more likely we will ourselves be eaten. The recognition of this truth is largely captured in the U. N.'s seventeen sustainable development goals. Even here in Goal 8 on economic growth and Goal 12 on responsible consumption and production, the limits that need to be recognized to achieve sustainability are rather fuzzy around the edges.

The dramatic conclusion of the July 2017 G-20 meeting in Hamburg, Germany, included a final communique that ended with 19 leaders of the world's largest economies stating that creating a sustainable environment was an absolute vital and top objective. This has been greatly undercut by the U. S. government standing alone in its commitment to withdraw from the Paris environmental accords and then withdrawing from the G-20 communique after the meeting was over and the communique signed. Such actions are both of substance as well as symbolic of the true differences of opinion on this vital issue for the right path forward for the human race.[11]

---

[11] Steve Erlanger, Alison Smale, Lisa Friedman, and Jule Hirschfeld, "Leaders Pledge Climate Action without Trump: 19 Members of the G-20 Adopt a Detailed Plan to Cut Emissions" *New York Times*, July 9, 2017, pp. A1 and A8.

Chapter 7
Where Is Technology Leading Us:
The Good, the Bad, and the Ugly

The lifeblood of tomorrow's world will be data, in all its manifestations. By developing machines that can finally capture and make sense of data we will unlock solutions to problems that have tormented us since the origin of man.

Shawn Dubravac, in Digital Destiny

The first ultra-intelligent machine is the last invention that man need ever make.

Irvin J. Good, British mathematician and close associate of Alan Turing

Introduction

Describing the future as the Cyber Revolution suggests that it could turn out to be a very scary place. And, as Carl Hiaasen suggested in his book *Assume the Worst,* this may very well be the best way to not be disappointed in the future.[1] The choice of the title for this book was deliberate. The word revolution was intended to suggest that the future is going to be characterized by much more than a few shifts in how we work and play. There is truly going to be incredible change. Some of that change is going to be scary bad. Other happenings may seemingly be half-crazy and disturbing. Yet not all hope should be abandoned. In fact, many other aspects of the impending future may very well turn out to be rewarding and positive.

The tools we now are seeing emerge, such as AI, smart robotics and ultimately the Singularity can and will provide amazing new capabilities. Peter Diamandis has voiced this optimism about how emerging artificial intelligence and new scientific and technological developments will bring forth a cornucopia of fresh opportunity. He is still among those who would love to go to Mars despite the fact that he is no longer a youngster. For instance, in his book *Abundance: The Future Is Better than You Think,* he repeats his favorite phrase that he was known to recite

---

[1] Carl Hiaasen, *Assume the Worst*, 2018, Alfred Knopf, New York.

J. N. Pelton, *Preparing for the Next Cyber Revolution*,
https://doi.org/10.1007/978-3-030-02137-5_7

often when we worked together in the early days of the International Space University: "The meek shall inherit the Earth, the rest of us will go to Mars." Diamandis has always been an eternal optimist about the world of the future being a brighter and better place to be.[2]

Similarly, Ray Kurzweil, in his book *The Singularity Is Near: When Humans Transcend Biology*, also exudes confidence that smarter and smarter machines can be a boon to humanity and a brighter future.[3] At the Arthur C. Clarke awards some 4 years ago he explained his optimism. He saw the future as being saved by new technology that allows us to combat climate change, bring solar energy and other recyclable energy systems to new levels of effectiveness and cost-efficiency and much more.

In his vision artificially intelligent machines with their ability to evolve intellectually and conceptually will become essential to the ultimate uplifting and future progress of the human race. It is this new ability to innovate in ever faster ways that will allow even bolder steps such as the opportunity to colonize and settle new worlds. The pathway to the future that is represented by the twenty-first century is likely to be bedecked with new opportunities and yet potentially oppressive ones as well.

The balance between new opportunity and potential threats to freedoms and even suppression of future rights is a delicate one. Balance will be everything in this future world of machine intelligence and potentially exponential evolution of machine intelligence.

It is important to remember that in almost all dimensions, the future is a one-way gate. Thus the best advice that one can offer is to assume the worst and seek for the best of all possible outcomes. Humanity has not progressed to where it is today by being defeatist. To be human is to possess a sense of optimism. We must always try our hardest to exploit the wonderful opportunities that the future will present and to overcome the adversity that is undoubtedly in store.

What we do know is that the future will be shaped by smart digital technologies and algorithms that will only get smarter and smarter. These new technologies are currently reshaping our world and bringing us the future filled with bucket-loads of change, industrial disruption and angst. Yet these same tools can provide us with a higher level of healthcare, longer life, much better education and training and more. Physical goods will be able to be produced to higher quality and so cost less. Learning systems will be available to us not only for the periods of childhood and young adulthood but for a lifetime. Lifelong learning will be essential in this time of constantly accelerating knowledge, exploding social change and a world that is a moving target.

But what about the angst and uncertainty that can already be anticipated?

Some of the greatest angst will come from the perception and perhaps the reality – of smart robotics taking over our jobs. Imagine a world in which workers can work not a 40-h work week but 168 h a week and 52 weeks a year. These willing workers would presumably not need a vacation, sick leave, pay raises, promotion,

---

[2] Peter Diamandis, *Abundance: The Future Is Better than You Think*.

[3] Ray Kurzweil, *The Singularity is Near: When Humans Transcend Biology*.

healthcare, or day care for their children. If union workers were ever concerned about scabs taking their jobs, then such smart robots seem like a nightmare. The idea that must emerge from this concern about the future is a more symbiotic relationship between humans and intelligent machines. The key to the future will be finding viable and useful human-machine interfaces. This will be critical in terms of redefining work effort, but also in terms of defining better ways to secure critical functions and infrastructure against cyberattacks.

As we know from Ray Kurzweil and neuro-scientist wizards such as Henry Markham, we are facing the prospect of super-computers married to AI algorithms that have the reasoning power of a human being perhaps as soon as 2030 – if not sooner. As Markham has described his quest in a TED talk, his goal is clearly defined. If only neuroscience would follow his lead, he insists, his Human Brain Project could simulate the functions of all 86 billion neurons in the human brain, and the 100 trillion connections that link them. And once that's done, once you've built a plug-and-play brain, anything is possible.[4]

But unfortunately the possible becomes an incredible array of things, both good and bad. There can be such good things as virtually free lifelong learning, low-cost health care, freedom from nine-to-five jobs and machines that can build everything from spaceships to custom-designed housing. On the other hand there could be cyberattacks against vital infrastructure, launch missile attacks or create a global pandemic of plague-like viruses, which human intervention cannot prevent.

A machine intelligence-directed war that is fought with chemical, biological or radiological weapons could be much more fearsome than World War II. Instead of tens of millions of people dying while fighting senseless and cruel trench warfare, or carrying out air bombing raids, the weapon systems to release virulent chemical, radiological or biological agents might be entirely under the control of machines.

It is hard to be an optimist and a pessimist all at once, but this may very well be what the future could present to us. How does one accentuate the positive and avoid the potential of a smart machine, under the control of a rogue or mad agent of destruction, disrupting modern society? How does one put a 'morality switch' in smart machines to ensure that they follow the first law of robotics as conceived by Isaac Asimov: "First do no harm to humans"?

Alternatives to Progressive Knowledge as Destined by Technological Man

Jacques Ellul, the French author of the book entitled *The Technological Man* posits the basic idea that the human species is defined by the search for new information and knowledge. We not only want to make tools, we want to make tools better and better. When it comes to warfare we want to make our arms deadlier and deadlier. If this is true, will not the smart machines we create end up wanting to do the same?

---

[4] Johnathan Keats, "The $1.3B Quest to Build a Supercomputer Replica of a Human Brain," *Wired*, May 14, 2013. https://www.wired.com/2013/05/neurologist-markam-human-brain/

Will not the smart algorithms we create be engendered to have a technological imperative for an ever-greater efficiency? How do we stop this new ability from being applied to warfare with inevitable disastrous effects? It is such concerns that have led thinkers, such as Elon Musk and Stephen Hawking, to suggest that artificial intelligence must be considered a potential threat to the longer-term survival of the human race.

Whether one considers the writings of Henry David Thoreau, Jacques Ellul, or even Arthur C. Clarke, the consequences of incessantly increasing technology end with the same concern. This is that either this technology could ultimately reach a stage where smart machines no longer consider humans necessary and rid themselves of their not very smart 'masters,' or even more likely, a mad or unprincipled person seeks to use this powerful technology to launch an attack on an opponent in such a way that a global disaster occurs.

Today, the current conflict between Western industrial society, along with its quest for technical advancement and scientific knowledge, stands in sharp contrast to the philosophic thoughts and religious ways of so-called Fundamentalist believers. These are those that seek enlightenment, religious insight, or traditional belief systems not driven by technology, or efficiency, but rather self-fulfillment. These religiously-guided people do not seek their guidance from science or empirical tests of reality. No, they seek insights and the "truth" from prophets who are able to commune with God. It is this basic conflict of technological pursuits versus religious or traditionally derived truths that are revealed via prayer and faith-based activities.

These diametrically opposed views of what guides human destiny and societal goals is today seemingly at the heart of many of the conflicts that rage around the world. Some believe that the pursuit of empirical science and technological advancement is the only path forward for humanity. Others argue that this represents the path to technological tyranny. Slavish attention to science and technological innovation is now seen by many millions, if not billions, of humans as anti-religious and opposed to their deep-seated cultural beliefs. In short, they reject so-called Western science as contrary to their fundamental religious beliefs. This split is sometimes oversimplified in contrasting Western industrial society with some belief systems found in Islamic, Sikh, or Zoroastrian, or remote village culture. This split can be seen between other divides as well, found in urban versus rural communities or Fundamentalist Christian communities versus Christians with a more humanist philosophic outlook.

The last few decades have only intensified the difference of opinion about whether technological advancement is a good or bad thing. Jacques Ellul, from a perspective shaped by the concepts and ideas expressed by Ralph Waldo Emerson and Henry David Thoreau, has written about what he saw as the deep failings of an economically and technologically driven world. He has said: "Modern technology has become a total phenomenon for civilization, the defining force of a new social order, in which efficiency is no longer an option but a necessity imposed on all human activity."[5]

---

[5] Fasching, Darrell (1981), *The Thought of Jacques Ellul: A Systematic Exposition,* Edwin Mellen Press, N.Y., p. 17.

There is a quite useful and insightful book by William Kuhns entitled *The New Post-Industrial Prophets: Interpretation of Technology*. In his book Kuhns has about half of the philosophical thinkers espousing the merits and advantages of modern science and technological wizardry and the other half against these. Thus he has the likes of R. Buckminster Fuller, Norbert Weiner and John von Neumann arguing the "pro" case and people like Siegfried Gideon and Jacques Ellul arguing the "con" viewpoint. What is interesting to consider is who among these insightful philosophers actually thinks that Technological Man has an option to stop in what seems to be a perpetual quest for new knowledge. We apparently must ask why and how things happen. Those who are espousing the positive case for technological advancement plus those who think that scientific advancement might be going to lead human society into harm's way all seem to intuit that we have little choice. Humans truly seem to be hell-bent in the quest for new understanding of the cosmos in which we exist.[6]

The fundamental question central to this chapter is exactly where technology is leading modern society? What is the relationship that is presumed to exist with regard to data, information, personal privacy, knowledge and wisdom, and what does empathy and morality have to do with these pursuits?

Even more important is the fundamental question as to whether a new understanding of concepts of physics, chemistry, biology, economics or politics will ultimately help to improve the human condition or not. Some, such as R. Buckminster Fuller, fervently believed that reasoning and heuristics that allow the effective processing and use of information are vital to progress and survival. His writings suggest that what was embedded in our souls, like a fundamental law of physics, was the essential drive to investigate and organize and understand – to overcome entropy in the universe that tends to diffuse order and run down to a lower energy state.

Yet there are those that strongly disagree with this assessment. They fear a future where smart machines can duplicate the capability of a human brain. This would represent the capability to operate at the 'exaflop' level of operations a second or 1 quintillion (10^{18}) operations in a single second. The idea of linking human thought with such machine processing power could produce many unwelcome results. One observer, known only as "Gray," who commented on this very proposal, has had this to say: "Only those without empathy are interested in merging or transforming the brain into a computer. In fact, it is already happening on a mass scale via technology and social media. It makes the human being predictable and controlled. How would you like for your every move to be predicted and controlled? Playing with these things are dangerous and eradicates free-will…."[7] (See Fig. 7.1).

The ultimate question about the human search for technological efficiency is whether this will become a threat to the human condition by demanding more and more efficiency and productivity. Can such pursuits drain the human soul and

---

[6] Walter Kuhns, *The New Post-Industrial Prophets: Interpretation of Technology* (1995).

[7] John Stuaghton, "The Human Brain vs. Super Computers—Which One Wins, Science ABC, May 2016 https://www.scienceabc.com/humans/the-human-brain-vs-supercomputers-which-one-wins.html

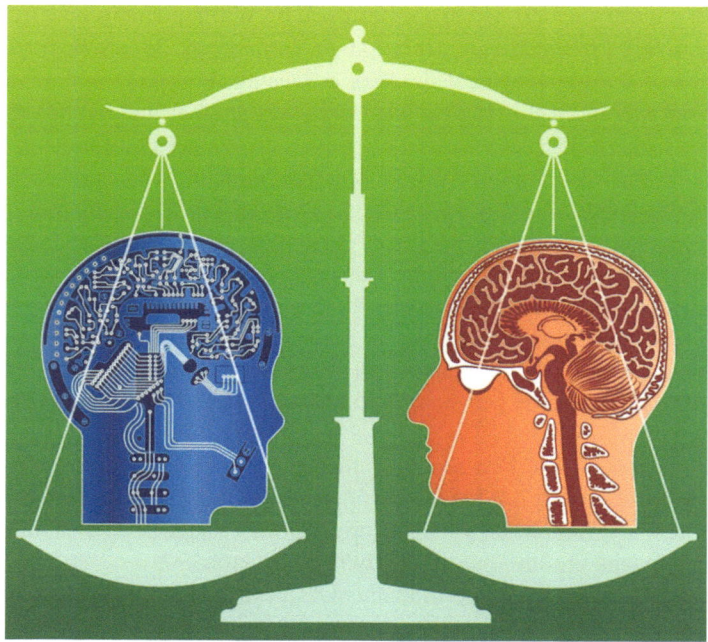

Fig. 7.1 Weighing the artificial vs. the human brain. (Graphic from the Global Commons)

diminish our ability to display morality and empathy? The end result, some would argue, would be to replace humans with more functional machines and presumably smarter and smarter artificial intelligence – i.e., the efficiency mandate.

What we do know is that today's economic systems are not fundamentally geared to produce more contented and productive people. No, the efficiency mandate of industrial systems is to create more rapid throughput and higher rates of information flows. Survival of the species and contentment are not a part of current mechanisms of capital markets, even though there are crusades that are seeking means to re-program capitalist markets in a new direction.

One estimate suggests that by 2030 there will be more than 30 billion smart machines capable of communicating data and perhaps preprocessing information that can also potentially be networked together. Other studies place the numbers much higher. The point is not to obtain the most exact and accurate forecast of net-worked smart devices. The fundamental issue is to recognize that the preponderance of information in modern society will be increasingly generated by machines and not people. Smart sensors, Supervisory Control and Data Acquisition (SCADA) systems, and appliances with application specific integrated circuits (ASICs) and other devices will churn out data 24 h a day 7 days a week.

Coping with Disruptive but Potentially Breakthrough Technologies and Concepts

The advent of truly capable artificial brains that can mimic the 86 billion neurons and trillions of linkages that can be made in the human brain at any instant will give rise to many questions. Some of these questions will be agonizing and difficult. There are some that are work-related or involve economic, political or social change. Others will involve ethics and morality. Finally some will be as basic as, what does it mean to be human and what does the survival of the human race ultimately mean? These questions are at the heart of the coming Cyber Revolution. The answers to these questions at this time are elusive, but the need to consider the implications of such questions is essential. It is not too early for political leadership to consider these profound questions and to hold hearings on where AI and intelligent machine capabilities will take us during the Cyber Revolution.

What Super-Automation Will Offer to an Overcrowded and Fast-Paced World

One of the first questions that should be asked is, why are smart machines being rapidly developed around the world? Are these super-fast AI algorithms and investigations by neuroscientists and super-computer engineers being pursued to make possible some of the many advantages that have been noted as possible in this new age of super-automation? The short answer is no. Not even close.

The prime motivation for the development of super-automation and ultra-fast artificial brain processing is industrial efficiency and cost cutting, corporate profits and "winning" within the context of international and capitalist competition. These goals are driving AI research and development activities and the push to achieve the Singularity forward – not altruistic objectives. The motivator is definitely not better education and healthcare, nor is it to relieve the toils of industrial or service workers.

At some point political, business, social and cultural leaders will need to figure out not only where our technological pursuits are leading us but where we wish human civilization to go and what our aspirations should be. Food for thought in this regard comes from the inspirational words of Eric Burgess, the author of *The Next Billion Years:*

> .....*the future of thinking beings is restricted only by their thinking, not by material laws. Even a plan to halt the growth of entropy in the physical universe is not beyond their ultimate capability. Earth has an astronomical future as a habitable planet of perhaps six billion years, enough time for a new race of thinking creatures to evolve from blue-green algae if all advanced terrestrial life forms become extinct. But the achievements that a continuously evolving intelligent species might make in 6 billion years are unimaginable.... But mankind has now entered a most critical period. He must deal wisely, both spiritually and materially, with the trauma of the next few decades if he is to inherit a future worth having.*[8]

---

[8] Eric Burgess, Introduction to *Global Communications Satellite Policy*, (1973) Mt. Airy, Maryland, Lomond Systems.

Coping with Cyberattacks and Techno-Terrorists

Almost anyone that reads the newspaper, listens to the news or gets updates on the web is aware that individuals, businesses, cities and even nations are subject to cyber attacks and even acts of techno-terrorism. The question is why, in the age of AI algorithms and proto-human brain power, must we be even more attuned to these types of attacks? The answer is, of course, that super-automation and automated industrial controls will increasingly permeate every aspect of our lives. In the age of the Singularity, the Internet of Everything will spread across the landscape of our lives. Our houses, our vehicles, our places of work, our vital infrastructure (i.e., the electrical grid, nuclear power stations, pipeline systems, water and sewage, etc.), satellite networks, military systems such as missile defense systems, and more will be automated and controlled by systems that could be attacked by viruses, malware and cyberattacks that could bring disaster to modern society as we know it.

This suggests one core thought. It is not a matter of stopping all possible cyberattacks. It means that we must use the latest in cyber knowledge and AI systems to prevent such attacks from having widespread effect, isolating the actions of bad actors and bringing them to justice. This means harnessing the new capabilities provided by proto-human artificial intelligence to design better human-machine interfaces to define aberrant behavior, shut down automated systems that could endanger human lives or create dangers for vital infrastructure. To put this effort into plain words, AI algorithms could be used to spot dangers and potentially hazardous actions and to put the brakes on or create shields against the malware spreading or halting a cyberattack from spreading. This, in effect, is both an effort to fight fire with fire but also to build in human-machine interfaces that are primarily designed to prevent cyberattacks if or when they occur.

Another way of putting this is to say that brakes of protection are increasingly more important than the economic push to create efficiency and higher throughput. Humanity, as it enters a new phase of evolution that we call the Fourth Wave economy, needs to recognize that we are truly on the threshold of a new age. In this new age, certain free market economy maxims have begun to become obsolete. As computer capabilities go from teraflop/sec. Speeds to petaflops/sec. and exaflops/sec., the increase in capability against human needs for efficiency and performance no longer have the payoff that was once possible. In the new age, protection against cyberattack and terrorism becomes a higher priority than industrial throughput or economic efficiency of production and labor-saving devices.

Lots of things are going to be different. Market-driven consumer economies that strive for higher throughput and better profit returns may still have a place in the world economy. Even so, there are now higher priorities. These priorities include the survival of Earth's biosphere, coping with climate change, and suppressing increasingly virulent super storms. These priorities will also include slowing exponential population growth and even creating planetary defenses against cosmic hazards such as asteroids and powerful solar storms that can wipe out our electrical grids and other vital infrastructure on which billions of people now depend. Yet another one of these new priorities would be the creation of new capabilities to halt the abuse of cyber technology – or at least put the brakes on those that would attack society via techno-terrorism.

Near the top of the list would be the creation of what could be called fail-safe human-machine interface systems designed to prevent the most virulent attacks against military systems, vital infrastructure and other smart systems on which we now depend. Closely tied to this need are ways to limit, halt and severely punish the dangerous exploitation of proto-human research capabilities. We need legal penalties and criminal prosecution systems with worldwide effect that can protect us against those who would attack society and modern civilization via the new AI capabilities that are now being created by the Henry Markhams and Raymond Kurzweils of the world.[9]

Co-existence with Smart Robotics and the Singularity

The future is not like the past. This is, of course, a truism. The world we live and work in has changed significantly over the years. There have been tremendous waves of change that came with the agricultural revolution, the Industrial Revolution and the post-industrial or services revolution, and it will change again with the Fourth Wave economy that is now evolving.

What is perhaps the most significant change is that humans will be sharing their world with smart machines. It is a sharing that is already happening, but this sharing is hidden below the surface of what we experience on a day-to-day basis. Our world is shared with supervisory control and data acquisition (SCADA) systems that control our elevators and escalator systems. They also control our water and sewage systems, our traffic light systems, our pipelines, our electric grids, and more. Today industrial control systems, automated temperature controls and smart sensors are already everywhere. The future of a world equipped with Rfids (Radio Frequency ID) tags for inventory control and the Internet of Things (IoTs) will mean a world that is automated and controlled by machine-to-machine (M2M) communications, and IT interactions with humans will be fewer and fewer.

Billions and soon trillions and trillions of smart devices and RFIDs will be able to talk to each other. These units will report on the health and status of airplane engines, washing machines, refrigerators, home and business security sensors, baby cams, automobile carburetors and more. Humans will become necessary only when there are alerts to a component that has failed or if the systems have been hacked in order to overload a particular Internet site with massive amounts of incoming messages.

But these are only the preliminary steps in the world of super-automation. The really big transition will only come after the Singularity has arrived and its effects

[9] Joseph N. Pelton and Indu B. Singh, *Digital Defense: A Primer in Cyber Security* (2015), Springer, NY.

are seen in the Fourth Wave economy. Around 2030, there will be AI-enabled processors that will be hard to distinguish from people if you are communicating with them by phone or Internet connection.

These smart processors can not only store and retain a great deal of information in their memory stores but can also reason like a human, talk like a human, and more. They can be designed to provide services like those of a professional accountant, an appraiser, or insurance or retail salesperson. These proto-human machines can be trained and equipped to undertake jobs now performed by millions, if not billions, of people around the world today. Whether it is teachers, professors, lawyers, doctors, dentists, chemical or structural engineers – no job, with the proper training and memory stores would be exempt from such proto-human processors, and the world is not prepared for this. We are not prepared economically, socially, culturally and certainly not politically. Even the business leaders who will calculate the economic advantages of automating their workforce with proto-human machines will find that they could be replaced.

The question then becomes one of how humans and proto-human smart processors can co-exist. The problem is that the world of automation, AI and the Singularity is quickly moving toward a time where far fewer humans will be needed to perform tasks and carry out work. At the same time the ranks of humanity continue to expand. In the arena of supply and demand, there are more and more people being produced at a time when there is less and less need for workers. This is, quite frankly, a big problem. We need people as consumers for an ever-expanding range of products and services. People need jobs to earn wages to purchase the goods, housing and utilities, but the whole idea of wages, jobs and expanded throughput of products and services makes less and less sense if work, products and services are going to be increasingly taken over by smart machines. Thus, the coming economic cycle of the Fourth Wave will involve sorting out a seeming mega-crunch dilemma. What do we do to cope with severely overcrowded cities, the adverse effects of too many people and shrinking levels of traditional employment opportunities?

It is important for economic, business, cultural and political leaders to start addressing these issues and concerns now. A number of possible reforms are necessary to address these issues.

The Re-invention of Public Policy for the Fourth Wave Economy

John Locke believed strongly that the invention of government and the creation of a system of laws, regulations and taxation were all created to protect 'natural' rights possessed by all humans. In the age of the Fourth Wave the only way that government can protect these natural rights, however, is not to protect individual property but indeed to save Earth as a livable habitat for humans. If Earth transforms into a fireball with highly pressurized gases like Venus there will be no property, no life and accordingly no liberty. So what can governments do to save Earth? In a word, plenty! (Fig. 7.2)

Locke's idea was that people should be willing to make a limited social contract with their government to give up their freedom and respect the sovereignty and power of their government as long as it ensured their life, liberty and protected their property. These ideas were encapsulated in the American Declaration of Independence written by Thomas Jefferson and are central to the basic tenets of democracies around the world. The twenty-first century dilemma is that the spread of technology, the growth of industrial systems and rise of human population and modern activities and services have now spread around the world. National governments can no longer ensure life, liberty and property rights because the entire world biosphere is at risk.

Nevertheless, national governments can still begin to act to help prevent the Anthropocene mass extinction. Further, the U. N. Sustainable Development Goals for 2030 can also help to postpone the worst dangers of climate change as the world heats up.[10] The problem is that these actions by national governments and the United Nations represent baby steps and avoid major reforms that are needed to truly reverse the dangers to the long-term sustainability of the world as a place for humans to survive for the longer term.

The question here is specifically what can AI systems, the Singularity and proto-human smart robotics do to help sustain Earth as a livable place for humans and all the other flora and fauna? The following is a listing of some of the ways where new AI systems could help save the world's ecology, bolster employment, improve

Fig. 7.2 John Locke's concepts of a social contract are still relevant today. (Image from the Global Commons)

[10] U.N. General Assembly, U. N.
un.org/sustainabledevelopment/s

healthcare and education, cope with overpopulation and help bridge the gap between developing and economically developed countries. This is not a complete list, but it is suggestive of the many positive contributions that might be possible:

- **New Employment Development.** For centuries now, human innovators and business people have been thinking of ways to do things better, cheaper and faster through the use of more efficient and smarter machines. There is some sense of ironic justice that now the time has come to have machines to help think of ways to find productive employment, tax incentives, or compensation systems that might assist humans live better lives. These new activities might involve: human-machine interfaces; globalization, environmental or archeological activities; bridging economic and cultural gaps; artistic, cultural, and intellectual pursuits; historic preservation; or more exotically, settlement of off-world sites or new subterranean, Arctic, or ocean-based enterprises.
- **Smart Energy Industries.** Accelerated evolution of sustainable energy systems and other new 'green' enterprises might not only create new jobs and enterprises but also help with the long-term viability of Earth as a habitable planet.
- **Innovations in Education and Healthcare.** As humans live longer and have less of a need to work in a traditional sense, the clear growth activity will be better health and educational programs that are available more universally around the world and throughout their lifetimes rather than conventional childhood and young adult programs. Here social interfaces between machine-based healthcare providers and learning systems and human doctors, nurses, instructors, and advisors will be a key part of this transition. The human touch will undoubtedly remain key from a social and cultural perspective if nothing else.
- **Taxation Systems.** The development of new taxation schemes can be aided by advanced analytic systems. There will be the ability to create alternative projections of the effects such systems might have to create different effects. These effects might be to reduce birth rates, increase the ability to bridge the gaps between people in different countries and across different ethnic and racial and linguistic classes, and make governments more responsive to the various needs of its citizenry.
- **Birth Control Innovations.** There is the possibility that proto-human computer intelligence could be given the task of addressing the issue of global overpopulation and urban crowding. Innovations here might help develop lower cost and easier-to-use birth control systems. It might also lead to more effective strategies to encourage and incentivize their use around the world. These strategies might be optimized down to a village or even the individual level. Exaflop/second systems might be able to customize incentives and strategies so that they are no longer geared to mass populations but be adjusted to the needs, concerns and aspirations of one person at a time.
- **Modeling and Forecasting.** In all of the above tasks, the power that will come with the Singularity will be the ability to undertake modeling and forecasting both on a macro and micro level. People in flood zones will be able to see what will happen to their homes with a 50-year flood event. Political and business

leaders, as well as the voters of a community, will be able to see the cumulative pollution results from a coal-fired electrical plan over three decades of operation in terms of emissions. The computing power that will be unlocked will be a powerful tool, not to reveal future trends but to see visualizations of pollution, erosion, and desertification events.

- **Protection against Cyberattacks.** This issue was already discussed in the previous section. It is perhaps one of the greatest challenges of our times not only to design software that can process information faster and automate more functions more efficiently but rather to develop AI software that can simply accomplish the protective result associated with Isaac Asimov's first law of robotics: "Do no harm to humans." We will need to enlist the best and brightest software to protect ourselves, our families, our communities and our countries against cyberattacks and to see that cyber criminals and cyber terrorists are arrested and go to jail.

Empathy, Morality and Ethics

The ultimate future of computer software and hardware development and creating faster and more effective AI systems will not be built on the basis of processing speeds or access to more bytes of memory. No, the ultimate artificial thinking machines will succeed – or fail – on the basis of capabilities in a different domain – the realm of empathy, morality, ethics and other aspects such as fault detection and the ability to perceive and stop such crimes as cyberattacks and techno-terrorist assaults on smart cities and national defense systems. One probably will have to start with a basic question. This question is, where or when will proto-human computer systems be able to think and reason like humans? This returns to the issue addressed earlier in the book, whether software systems can be sued in a court of law for exercising 'poor judgment' or insufficient concern for human life? In short, in the Fourth Wave world, will AI systems exercise judgment and logic? Can smart machines truly demonstrate such human aspects as empathy, morality or ethical behavior?

In the world of digital processing, there can only be 0s or 1s. There is no option other than these two values. With chaos theory and 'fuzzy logic' systems there can be a wide range of shades of gray. We need more than 0s and 1s to describe our world. We need a range of options to consider all the conflicting values that need to be considered in our complex future world. We are still a ways away from such issues coming to the fore. But the days when such issues can and will be asked in a court of law, or at least in the court of public opinion, are much closer than most people think. Will we see the marriage of digital logic systems and fuzzy logic systems in the not too distant future? We already know that for some automated functions, such as backing a truck into a narrow passageway, can be done far more effectively using 'fuzzy logic' systems as opposed to digital processing algorithms.

Conclusions

There is Pelton's Law of Prediction. This law is not to make a 5-year prediction because a legal, regulatory, social or cultural complication will slow the process down and complicate matters. Thus such a forecast development will not occur in a timely manner. Never make a 10-year prediction because technology is moving so very fast you will find a new scientific, engineering or process development has been overlooked or not anticipated. This will prove you wrong. Thus always make a long-range forecast of 25–50 years into the future so that no one will be around to check up on you, or you will be long deceased, so you will not care whether you were right or wrong.

It would be useful if the Fourth Wave, the Singularity and the age of proto-human computer systems and smart robotics were safely 25–50 years away from happening. Thus it would be relatively safe to make a prediction about how this all plays out. In this best of all possible worlds, people would live much longer and enjoy contented lives amid a future of excellent education and healthcare systems that are made available at low cost. There would be world peace and global harmony. In this world, there would be a relatively stable global population of only five to six billion people, and problems with global climate change, overcrowded cities, the digital divide and technological under-employment would have all been faced intelligently. All of these problems would be largely solved.

We would have even started to colonize the Moon and Mars, built a solar shield at L-1 to protect against destructive coronal mass ejections and put in place an asteroid circling the Moon to serve as a sort of protective billiard ball to deflect a stray killer asteroid from smashing into Earth.

The problem is that many of the problems discussed in this chapter will emerge as serious challenges within the coming decade or so. Few, if any, of the above predictions would describe a future reality. The real prediction would be serious overpopulation, continued rise in global temperatures, more and more violent storms, serious climate change issues, under-employment and even further economic and First World versus Third World stratification.

The only real hope is that we really can deploy our new cyber technology in such a way that economic, business, and political leaders of the world embrace new ideas on creating new jobs in the sustainable energy sector, finding a way to reduce global population expansion and using our smart machines to try to move toward a new world of abundance. Super-automation truly is a challenge, but smart machines can truly help us find new and better answers.

Chapter 8
Healthcare and Educational Systems in the Age of the Cyber Revolution

*Achieving inclusive and quality education for all reaffirms the
belief that education is one of the most powerful and proven
vehicles for sustainable development. This goal ensures that all
girls and boys complete free primary and secondary schooling
by 2030. It also aims to provide equal access to affordable
vocational training, to eliminate gender and wealth disparities,
and achieve universal access to a quality higher education.*

U. N. Sustainable Development Goal 4 for Quality Education

*We have made huge strides in reducing child mortality,
improving maternal health and fighting HIV/AIDS, malaria and
other diseases. Since 1990, there has been an over 50% decline
in preventable child deaths globally. Maternal mortality also
fell by 45% worldwide…. Despite this incredible progress, more
than 6 million children still die before their fifth birthday every
year. Sixteen thousand children die each day from preventable
diseases such as measles and tuberculosis.*

*U. N. Sustainable Development Goal 3 for Good Health and
Well-Being*

Introduction

The world as we know is changing – politically, economically, socially, environmentally, and especially technically. On the economic front, many nations have
faced rising deficits.

Nation after nation, from Spain to Greece, from Eastern Europe to across Africa
and South and Central America – the problems associated with budget deficits and
eroding value of national currencies have spread globally. Even economically stable
governments, such as the United States, have passed tax reform that is projected to
increase U. S. governmental deficits by trillions of dollars in coming years.
Environmental concerns associated with climate change, rising pollution levels and
population growth continue to increase across the planet. Others have suggested that

© Springer Nature Switzerland AG 2019
J. N. Pelton, *Preparing for the Next Cyber Revolution*,
https://doi.org/10.1007/978-3-030-02137-5_8

economic systems driven by capitalist economic expansion and the incessant pursuit of profit at all costs are a threat to the environment, to economic stability and even democracy.[1]

More and more governments have recognized they cannot conduct business as usual without courting bankruptcy. In 2008 the largest global recession since the Great Depression of the 1930s began and lasted for almost 6 years, resulting in more widespread recognition that many of the social policies must be rethought. With a population where 80-year-olds are growing apace, universal retirement at age 55 in countries such as France and retirement of teachers and federal workers after 30 years of service, such as in the United States, policies such as these are simply not viable. Virtually free educational and medical services for an ever-growing percentage of the population without dedicated tax revenues is not economically feasible in either a capitalist or socialist society. If people live longer, then many would argue that older retirement ages must ensue. If medical care costs escalate, then either one must find a way to cut costs, reduce benefits or fund supplemental retirement and medical insurance, either individually or on a group basis.

Many have contended that one of the keys to the future and rising health and education costs is technological innovation. New technology has already disrupted the hotel industry (i.e., Airbnb), the taxi industry (i.e., Uber and Lyft), and the brokerage business (i.e., Schwab and Fidelity). One of the key questions of the day is whether the time for major technological reform has now come for the health and education sectors as well. Can disruptive telecommunications and AI technologies help to make health and education services that are more cost-efficient? An even more fundamental question is, can they make these essential services more responsive to the needs of citizens and students?[2]

Proponents of tele-education and tele-healthcare argue that, in time, they can make these services more efficient, more proficient, and more cost effective. IBM's development of its prodigious Watson super computer, loaded with exabytes of information, and especially its Watson system devoted entirely to medical information and healthcare, are among the new tools that are being developed to support tele-health-based services. Let's look at the current situation.

The United States, with dubious distinction, leads the world in its spending on education and health care. It is now spending about 9–10% of its GDP on education, not including extensive training programs, and about 16–18% of GDP on healthcare. The insult to injury aspect of this statistic is that the results are clearly worse than in nations that spend far less of their money on these services. The students from many of the OECD countries score higher on standardized test scores than the United States, while spending less than what is spent in America. Likewise, there are many countries where their citizens live longer and where there are fewer childbirth deaths than in the United States, despite spending far more.

---

[1] Robert Kuttner, "Can Democracy Survive Capitalism?" (2018) W.W. Norton, New York.
[2] David von Drehle, "Going to school no longer means going to school", Washington Post, May 16, 2018, P. A15.

Table 8.1 OECD data on key healthcare statistics

Country	Life expectancy	Health care costs/cap ($ US)	Health care costs as % of GDP	% of gov't funds going to health care	% of total health care paid by gov't
Australia	81.4	3137	8.7	17.7	67.7
Canada	80.7	3895	10.1	16.7	69.8
France	81.0	3601	11.0	14.2	79.0
Germany	79.8	3588	10.4	17.6	76.9
Japan	82.6	2581	8.1	16.8	81.3
Norway	80.0	5910	9.0	17.9	83.6
Sweden	81.0	3323	9.1	13.6	81.7
United Kingdom	79.1	2992	8.4	15.8	81.7
USA	**78.1**	**7290**	**16.0**	**18.5**	**45.4**

OECD Health Care Data 2010, http://www.oecd.org/dataoecd/

The following tables (see Tables 8.1 and 8.2) suggest that the United States has a lot to learn from other countries in both fields of education and healthcare. These tables also indirectly suggest the wisdom of developing and applying better and "smarter" technology in the healthcare and education fields in addition to carrying out legislative and structural reforms. This is particularly the case in the United States. The alternative is to see an increasing percentage of students and elderly suffering from substandard education and healthcare and a lower and lower return on the huge amounts of capital and operating costs invested in these fields.

Table 8.1 shows that U. S. life expectancy falls below eight other OECD countries even though its healthcare costs are sometimes double that of these other countries, and healthcare consumes a larger share of the U. S. Gross Domestic Product. The U. S. federal government also spends a higher total of its national budget on healthcare (i.e., 18.5%) even though this money represents only about 45% of the total cost of healthcare. These figures add up to what could be considered a "lose, lose, lose" situation for the United States

If one suggests that longevity is not necessarily a fair measure of healthcare, one could also opt for infant mortality rates. Here, too, the United States lags behind these other countries with a mortality rate of 6.5 per thousand compared to about 5–6 per thousand for these other countries.

Educational statistics as complied by the OECD and other sources again shows that the United States spends more on its educational programs but with poorer results. Table 8.2 shows that the United States lags far behind other countries when it comes to comparative test results. Although Table 8.2 shows results in math, science test results also show a similar pattern. Among OECD countries, only the Republic of Korea and Iceland spend on a per capita basis a comparable amount of its GDP on education to the United States. The United States, certainly in absolute terms, outspends all countries. Even in terms of a percentage of its GDP, the United States outspends in descending order: Denmark, Canada, Sweden, New Zealand, Belgium, Slovenia, France, Switzerland, United Kingdom, Finland, Mexico,

Table 8.2 U. S. student math scores compared to other nations, Grades 4, 8 and 12

	Grade 4		Grade 8		Grade 12	
Rank	Nation	Score	Nation	Score	Nation	Score
1.	Singapore	625	Singapore	643	Netherlands	560
2.	Korea	611	Korea	607	Sweden	552
3.	Japan	597	Japan	605	Denmark	547
4.	Hong Kong	587	Hong Kong	588	Switzerland	540
5.	Netherlands	577	Belgium	565	Iceland	534
6.	Czech Republic	567	Czech Republic	564	Norway	528
7.	Austria	559	Slovak Republic	547	France	523
8.	Slovenia	552	Switzerland	545	New Zealand	522
9.	Ireland	550	Netherlands	541	Australia	522
10.	Hungary	548	Slovenia	541	Canada	519
11.	Australia	546	Bulgaria	540	Austria	518
12.	United States	545	Austria	539	Slovenia	512
13.	Canada	532	France	538	Germany	495
14.	Israel	531	Hungary	537	Hungary	483
15.	Latvia	525	Russian Fed.	535	Italy	476
16.	Scotland	520	Australia	530	Russian Fed.	471
17.	England	513	Ireland	527	Lithuania	469
18.	Cyprus	502	Canada	527	Czech Republic	466
19.	Norway	502	Belgium	526	United States	461
20.	New Zealand	499	Sweden	519	Cyprus	446
21.	Greece	492	Thailand	522	South Africa	356
22.	Thailand	490	Israel	522		
23.	Portugal	475	Germany	509		
24.	Iceland	474	New Zealand	508		
25.	Iran	429	England	506		
26.	Kuwait	400	Norway	503		
27.			Denmark	502		
28.			United States	500		

"Math scores Academic Failure-International Test Score Results", http://4brevard.com/choice/inter-national-test-scores.htm. A similar pattern can be seen in science test scores with American students slipping in performance against students in other countries as the progress from grades 4 to 12

Australia, Poland, Chile, Hungary, the Netherlands, Portugal, Austria, Norway, Japan, Italy, Brazil, Czech Republic, Germany, Ireland, Spain, the Slovak Republic, the Russian Federation and Turkey.[3]

The future trend lines for healthcare and education are ominous. People are living longer and thus some need healthcare even past 100 years of age. Medical research is expanding both medical knowledge and the amount of knowledge doctors, nurse practitioners, and nurses must acquire. The cost of medical care and drugs is outstripping inflation. The number of doctors per capita is shrinking, especially in rural and remote areas. Education has similarly adverse trend lines.

The number of people to be educated in the twenty-first century is larger than that represented by all the people requiring education since ancient times, when formal teaching began. The amount of new information being added the global information database is now expanding by at least 4 exabytes a year. That is 4,000,000,000,000,000,000 bytes of information (or the equivalent of several quintillion words). The information added in just 1 year is equivalent to many millions of times the amount of information known in ancient Greek, Roman, Middle Eastern and Chinese society. The amount of information that teachers and students are asked to know and assimilate has mushroomed in just the last century and continues to expand exponentially. Again, in contrast, the number of teachers per capita is decreasing.

The bottom line is that expert systems, artificial intelligence, the Internet of Everything and eventually self-aware machines of the age of super-automation will need to assume a greater and greater role in both healthcare and education to meet the demands of a twenty-first century world. Computer-assisted education, self-directed education using video labs, and computerized learning systems are likely to be a significant part of the future of education. Likewise these techniques will need to be applied to the training and education of doctors, diagnosticians, nurse practitioners, and nurses. Already there are sophisticated tele-health programs that deliver increasingly high quality healthcare to rural and remote areas, and the trend will undoubtedly continue. Part of the problem comes from information systems that process and store information inefficiently and inefficiencies in healthcare insurance companies and governmental support agencies. Other problems arise between the confusion that exists in many countries between what is, in fact, "child tending" for working people and what is education.

What is the future of healthcare and education in the United States and OECD countries, and what is the future of healthcare and education for the rest of the world? The answer is far from clear, but one thing is sure. Almost everything will be different. These services will be different in terms how they are delivered. They will be different in terms of who delivers them. Certainly there will be a big difference in how people or sentient bots are trained or designed to perform. They will be different in terms of cost structure, governmental agency organization, and industries that support these service industries.

---

[3] "Education at a Glance: OECD Indicators: Chapter B: Financial and Human Resources Invested in Education." http://www.oecd.org/dataoecd/

Spending Choices for Social Services Versus National Defense and Infrastructure

How much money should national governments spend and what functions should they perform? How much on national defense, and building roads, dams, and vital infrastructure? How much on education and healthcare? How much on research and development, and housing and social services? What should be allocated for retirement benefits programs? The allocation of funds is in many cases defined by quantifiable need and logic, and in other cases by emotional response and political clout.

The top priority for most national governments is to provide for national defense. But the meaning of national defense has changed significantly over time. Today, national defense covers at least the armed forces (navies, air forces and missile systems); intelligence and surveillance systems and satellite networks; disaster recovery and response; and, most recently, cybersecurity and the prevention of terrorist attacks. Clearly the role of government has expanded over time, and technology, increasing military and social needs have driven up the cost of government.

One of the tougher questions facing politicians and legislators is that of how much should a national government spend on social services? Should they be funded and managed at the national level of government or by provincial or local governments where perhaps needs are much better understood? How much should national governments spend on health care and/or on education and training? The role that governments now play in various sectors and the allocation of money for various functions clearly varies from country to country.

Allocation of funds and national government expenditures can in some cases be defined by a single event and control funding for decades to come.

The September 11, 2001, attack by aircraft on the World Trade towers and the Pentagon was such a defining event for the United States. In this tragedy about 3000 people lost their lives. This one event led to the creation of the Department of Homeland Security (DHS) and the annual spending of additional billions of dollars on various types of security measures. These range from security checks at airports to elaborate efforts to seek intelligence about the activities of terrorist groups. There could be questions as to the logic of such extreme protective strategies in the context of the actual threat.

For instance, one might compare the billions of dollars spent on security-related expenditures allocated for anti-terrorism activities versus the tens of millions spent for highway and road safety. Road fatalities in America claim the lives of some 40,000 people each year on U. S. roads and some three million are injured.[4] Or one could also consider deaths by firearms in the United States. Between 1968 and 2018 some 1.5 million people had been killed by firearms. In contrast, if one adds up all of the war fatalities experienced by U. S. soldiers in all of the wars from the Revolutionary War through the Iraq war the entire total comes to 1.2 million.[5]

---

[4]Neal Boudette, U. S. Traffic Deaths Rise for a Second Straight Year, *New York Times,* Feb. 15, 2017. https://www.nytimes.com/2017/02/15/business/highway-traffic-safety.html

[5]Guns in the US: The statistics behind the violence, BBC, January 5, 2016 http://www.bbc.com/

In the area of education and healthcare, there is debate as to whether this is something that national governments should be involved in at all – at least in the United States. There is perhaps some thought that there might be some funding of research and development of new medical or educational technology and systems in the United States, but that local and state/local government should be providing for schools and funding the salary of teachers and administrators. But just as the 9/11 attack redefined U. S. governmental activities and funding related to terrorist attacks, intelligence, surveillance and cyber defense, the breakthroughs in AI systems and the Singularity could redefine the future of education, medical and healthcare services. If AI capabilities can significantly increase the quality of education and healthcare and significantly reduce costs, this could redefine almost every aspect of how these services are rendered and extend these service offerings to virtually universal coverage around the globe. Limitations that exist today could be virtually eliminated, unless constrained by the market externalities so brilliantly identified by Adam Smith many centuries ago.

The Coming Revolution in Smart Health Services

The revolution in smart health and medical services started decades ago when isolated areas in Canada, Australia, the United States and island countries in the Caribbean and South Pacific began to develop tele-medicine and tele-health services, beginning even in the 1950s and 1960s. By the time of the Intelsat Project Share (Satellites for Health and Rural Education) in the mid to late 1980s, the idea of tele-health and tele-education services had already been well established. The University of the South Pacific and the University of the West Indies had found innovative ways to share faculty, medical services and healthcare advice among the islands covered in their network tied together by satellite and radio telephone services. Likewise, Canadian and Australian programs combined remote medical access by doctors and nurses with telecommunications and data links for training and diagnostic purposes.

Tele-health programs that provide instruction on basics such as nutrition, coping with infant diarrhea and instruction on how to set a broken arm or leg are now common using satellite and Internet connections across the globe. Under Project Share, the Miami Children's Hospital arranged for a medical training session by world leading medical experts on AIDS treatment. This was linked via satellite connections to enable the participation by over 50,000 doctors and nurses via Intelsat satellite connection for interactive viewers in North and South America, Europe and Africa.[6]

Today over a million people are connected via the INSAT program in India to receive remote education, training and medical treatment and healthcare advice. In China, the National TV Satellite System connects more than 10 million people in

news/world-us-canada-34996604

[6] Joseph N. Pelton, John Howkins, and Jim Stevenson, *Project Share* (1989) Intelsat, Washington, D. C.

remote parts of the country. This program started with only a few dozen sites under Project Share during the 1986–87 period and expanded to 90,000 television receive-only stations (See Fig. 8.1).

The growth of Internet broadband and remote cable, wireless and satellite coverage has expanded such programs on all the continents of the world, including Antarctica. Telesat Canada has been one of the leaders in tele-health services, but many of the low-Earth orbit satellite networks, such as those planned to provide truly global coverage for broadband Internet, will enable Internet-based health and education services worldwide at a very modest cost. The issue of the future will less the ability to achieve broadband connectivity; rather the problem will be availability of educational programming, online access to doctors, consulting nurses, and medical technicians. It is in this regard that AI-based systems and systems like I.B.M.'s Watson for medicine that can serve to provide global access to those that today lack the ability to connect to trained doctors and nurses. This involves more than remote access to an accurate medical diagnosis, but also access to drugs and medicines and medical practitioners. Already we are seeing the development of bots that are capable of remote treatment and even sophisticated treatments such as remote surgery and dental care. These capabilities were first developed for the treatment of astronauts or researchers in Antarctica or in the ocean, but in time they will expand to treatment in remote areas.

The Virtual Incision Company in Lincoln, Nebraska, has, for instance, developed fist-sized robotic devices that weigh just 0.4 kg (or about 9/10th of a pound).

Fig. 8.1 The Project SHARE locations in China for tele-education and tele-health

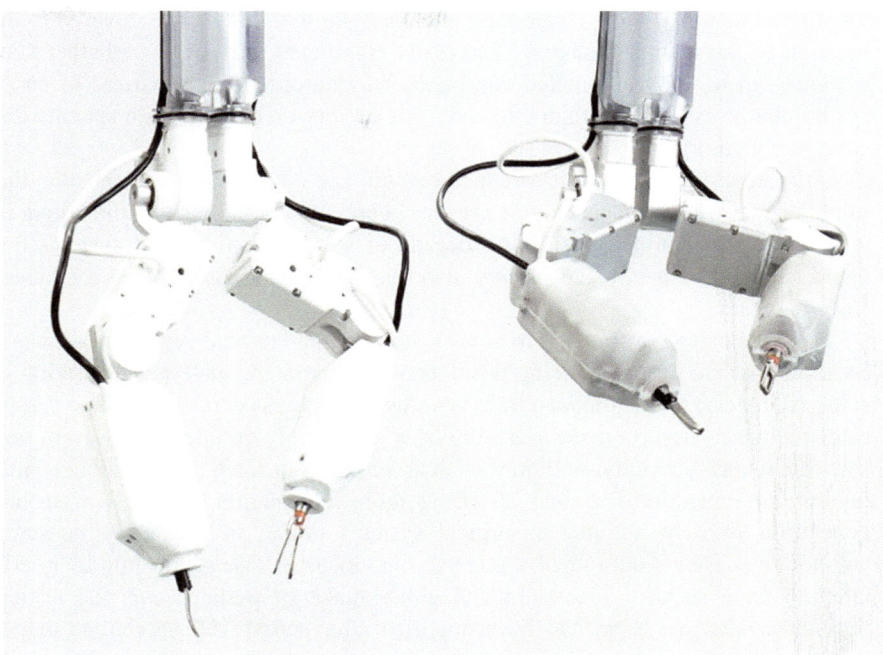

Fig. 8.2 Robotic surgery is under rapid development. (Graphic courtesy of Virtual Incision Company of Lincoln, Nebraska)

These are small enough that they can be introduced into a human cavity swelled by gas in order to perform surgery. Such robotic tools can operate on an ailing colon, remove an appendix, or repair a gastric ulcer via remote controllers. This system can cut, cauterize and perform other surgical procedures and has been tested on the International Space Station with simulations but not with actual surgery. On the ground, these systems have operated on actual pigs. The next step, currently in process, is to design and develop systems that do not have surgeons at the controls but would be fully robotic with AI in charge. Clearly fully automated systems would be necessary to operate on astronauts further from Earth orbit or early settlers on the Moon or Mars because of the time delay associated with great distances[7] (See Fig. 8.2).

Small-scale, less invasive robotic surgery with miniature systems are already here. The da Vinci surgical system, developed by Intuitive Surgical, has robotic systems capable of carrying out non-invasive surgery. Their website indicates that some five million surgeries have been conducted using this system, one every 60 s. Further, this system is also available for diagnostic purposes. Key to the operation of this system is a three-dimensional, high-definition visioning system designed to

---

[7] Aviva Rutkin, "Mini Robot Space Surgeon to Climb inside Astronauts" New Scientist, April 1, 2014 https://www.newscientist.com/article/dn25341-mini-robot-space-surgeon -to-climb-inside-astronauts/

provide a magnified view inside the patient's body that can be maneuvered to the location of the intended surgery. The really significant next step is whether such systems can be fully automated and made truly autonomous by virtual of an AI system that is essentially trained to carry out surgery on patients with specific diseases or failed organs.[8]

Ultimately, the question is whether we will see AI-enabled medical bots that might provide universal care services everywhere. The thought is that this might be possible in the coming years, but a decade or so ago, miniaturized systems that could carry out non-invasive surgery also seemed to be as far away as a Chicago Cubs pennant.

There are many questions as to how an autonomous surgical system might work. Systems that are currently being developed and operated, such as self-driving or self-parking cars and automated train systems, as expert systems that follow certain rules such as to stop for pedestrians or obstacles and rely on information from sensors to navigate correctly. Activities such as surgeries or medical operations would involve the collection and analysis of far more information. The first or second generation of such automated surgical systems might, in fact, only be semi-autonomous. They would involve a consulting doctor whose role would be to collaborate on a surgical procedure and either make or perhaps override critical decisions. After a system had been qualified after perhaps 50 operations then it might operate on its own without having to have its actions endorsed or supervised by an attending physician.

The evolution from expert systems that follows rules of practice from experts in the field to more complex and human like analysis will come over time. This will see the development of higher and higher levels of artificial intelligence. These will be systems that are able to assess past experience and extensive information from its memory banks and then make judgments about probabilities and possible consequences.

The point is not only will artificial intelligence and automated logic systems become more capable with higher and higher skill levels and competencies in many fields, but these systems will in time be able to be both mass produced and also improved in their competencies. The bottom line is that with healthcare and medical diagnosis, the cost of these automated systems with AI will become more and more affordable and their services more pervasive.

So, what sort of implementation model will be followed? Will this follow, for instance, the model of pharmaceutical companies? These companies develop new drugs and then use patents to maintain high prices for their medications for many years. Or will some other model apply? The international competitive model that applies to computers and software suggests that the cost of super-automated and AI-controlled software in the age of proto-brain bots should see a revolutionary impact on consumer prices. Unless market externalities intervene to distort prices our world should dramatically change. We should see plummeting costs and prices for all types of services, from surgical operations to accounting, from university education to legal fees.

---

[8] Intuitive Surgical Products, Accessed May 12, 2018. https://www.intuitivesurgical.com/products

Indeed it may well be that the significant economic thrust in the world today could well be the trend toward open software systems. The Hadoop Distributed File System, released by Yahoo in 2007 as an open software scalable file management tool, was perhaps a key precedent for the future. This massively scalable system is now critical to such operations as Amazon's worldwide ordering system. This type of action by Yahoo systems engineers may have been the start of a revolution that will eventually shake world markets for super-automated AI services in the years and decades to come.[9]

The Future of Educational and Training Systems in Age of the Cyber Revolution

Education and healthcare systems are often seen as closely linked. These are vital services that governments typically provide and are for the most part seen as universally needed across the planet. Trained practitioners in specialized facilities provide these services, and tests are administered in order to monitor progress and status of patients and students. Programs such as India's INSAT program are designed to provide health and medical care and educational instruction.

The long-term perspective on tele-medicine and tele-educational services is that if they can be effectively provided to rural and remote areas then other tele-services can also be efficient and cost effective. Indeed Hughes Network Systems, which designed satellite networks to provide rural tele-education and tele-health services, have also provided other capabilities such as tele-banking and remote governmental services. In Indonesia, the Palapa Satellite System was designed to provide IT and communications services to remote mining and energy sites along with tele-education and tele-health services.

Tele-health services are generally more demanding than tele-education services because medical diagnosis requires more precise and broader band capabilities to conduct precise and accurate scans, imaging and patient monitoring. The latest systems to provide broadband Internet services in remote locations, such as the large-scale low Earth orbit constellations of One Web and the Telesat Canada constellation, and proposed systems by such groups as Space X, Boeing, Leosat, Facebook, the Chinese Hongyan constellation, etc., have sufficient bandwidth ability to support educational radio and television and Internet streaming, as well as remote tele-health services.[10]

The problems with the economic and widespread consumer access to these systems are not with the communication networks bandwidth. Further super-automation capabilities and AI-based systems will support a wide range of education services.

---

[9] The Hadoop Ecosystem, Last accessed May 12, 2018. https://www.thinkbiganalytics.com/leading_big_data_technologies/hadoop/

[10] Larry Press, "Important Developments on Low-Earth Orbit Satellite Internet Service (2017 Review)" CircleID, Jan. 2, 2018. http://www.circleid.com/posts/20180102_important_developments_on_low_earth_orbit_satellite_internet/

No, the problem with access will be issues such as the high cost of ground equipment for tele-education and/or tele-health services and tariffs imposed on this equipment and/or governmental limitations and tariffing requirements associated with accessing these new types of services. Groups such as Geeks without Frontiers have developed white papers to promote the cause of low-cost access to broadband Internet and Internet-optimized satellite networks that are designed to bring lower cost tele-education and tele-health services to developing countries – particularly in Asia, Australasia and the South Pacific, Africa, South and Central America and the Caribbean.

Geeks without Frontiers co-founder Michael Potter said at their Thought Leadership Conference held in October 2017 that they hoped to bring low-cost broadband communications to an additional one billion people in the decade ahead. Specifically he said: "The Community Connect vision is to enable 100% availability of broadband communications services everywhere, providing businesses, governments, hospitals, schools, NGO's, individuals and others with access to broadband services, wherever they are located. This will help to bring the educational, healthcare, social, economic and e-government benefits to communities everywhere and facilitate and accelerate the achievement of the U. N.'s Sustainable Development Goals (SDG's)."[11]

The problem with these initiatives to some leaders in nations with developing economies is that all of these systems tend to be originating from overseas and offering tele-services that are in no way home-grown. The national political and economic leaders into these new networks don't always buy into the education and health care programming and tele-services that they are offered. The initial economic failures and bankruptcies of the Iridium and Globalstar satellite constellations in the late 1990s came in part because of problems with national landing rights, revenue sharing formulas between the satellite service providers and the local telecommunications companies and the high cost of consumer units to connect to the networks, especially after the high import fees and taxes were imposed on the satellite ground access terminals. The low cost and AI-enhanced software for education and healthcare that can flow through these networks are not likely to ease the concerns that these systems benefit developed countries more than developing countries.

The Afristar satellite system that was designed to bring low cost satellite radio services to Africa for education and health care as well as commercial radio channels was unsuccessful, in part, because the governments across Africa tended to impose tariffs and import duties that amount to 100% of cost of the small compact receiving units. What started out to be $50 receiving units became $100. And there were other challenges. These included problems such as reliable power supplies in

---

[11] "Geeks Without Frontiers Releases Its 'Community Connect' Global Broadband Initiative at the Geeks "Connectivity is the Revolution!" Thought Leadership Forum" GlobeNewsWire, Oct. 19, 2017.http://markets.businessinsider.com/news/stocks/geeks-without-frontiers-releases-its-community-connect-global-broadband-initiative-at-the-geeks-connectivity-is-the-revolution-thought-leadership-forum-1004899647

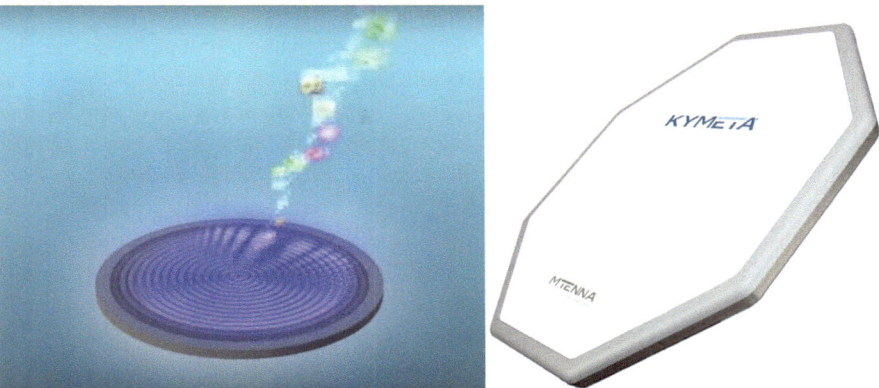

Fig. 8.3 Electronic tracking beam and newly manufactured Kymeta flat antenna. (Graphics courtesy of Kymeta)

remote locations that either were not available or were subject to spiking that could burn out these units as well. The best-intentioned systems can go astray if the local issues, concerns, facilities and revenue constraints are not taken into account. It is one thing to conceive of a low cost, globally available broadband streamed Internet service for education and healthcare that might be provided via new satellite constellations and smart software systems; achieving a successful implementation of these new systems is quite another thing.

There is one new technology that can help satellite constellations in low Earth orbit to operate much more cost effectively. This new technology involves the use of Earth stations that do not have to physically track the fast-moving low Earth orbit satellites as they pass overhead and are visible for only a few minutes. In fact they need to typically switch from satellite beam to satellite beam about once a minute. These new ground antennas use newly developed metamaterials to create electronically steered beams rather than requiring physically steered dishes. These flat beams with electronic steering have just come to market, and the current market leader is a firm known as Kymeta. This firm is backed, in part, by none other than Bill Gates, the co-founder of Microsoft and one of the early backers in the 1990s of a low Earth orbit constellation known as Teledesic[12] (See Fig. 8.3).

The technologists always tend to assume that the best shiny new technology will provide the solution to problems. There was a time when the London City Council thought that automobiles would save them from the huge environmental problem of their day, which was horse manure on their streets. Thus it is always key to remember that science and technology is only a part of the answer and a part of the equation. Economics, political distortions of the market and social and cultural acceptance of new solutions are often the bigger part of any new innovation in how a key service is offered and accepted within any society.

---

[12] Kymeta, "Connectivity where it has never been before" Last accessed May 13, 2018. https://www.kymetacorp.com/

Too often in today's world, scientists and engineers only talk to each other. Business people talk only to business people. Economists, politicians and regulators also tend to form closed loops. Teachers talk to teachers and students. Religious and cultural leaders also tend to have their own clubs. There are in this world of some 7.5 billion people silos within which information is communicated, but there is not a good and useful exchange between these groups. Further there tend to be networks that are largely restricted to a national discourse. The modern world is increasingly based on world trade and more and more dependent on finding solutions to global problems. These include such problems such as overpopulation, pollution, super-automation, under-employment and technology displacement, dirty energy consumption systems, global warming, severe weather patterns plus massive storms and climate change. Newer and more effective education and healthcare systems can help to cope with these problems, but only if there is better and more interdisciplinary communications and information sharing.

Conclusions

The speed with which new technology and tele-services are being developed in today's world is stunning. And, in fact, the rate of acceleration of these changes is increasing. New systems of broadband communications – both mobile and fixed networks – are expanding across the globe. New smart bots are taking on new tasks. And there are even systems being developed to distinguish whether someone using the Internet is a person or a bot. A recent issue that has emerged as a serious legislative concern is whether consumers should be formally and officially notified whether they are communicating with a live person or a bot.[13]

The dramatic expansion in global broadband connectivity and the accelerating expansion of AI capabilities are redefining what is possible today. There are massively scalable computer file systems such as Hadoop. There are researchers claiming to be able to create the equivalent of a fully functional human brain in terms of processing speed, memory and simulated synapse connections within 10 years or so. This is what we call the artificial proto-brain that could very well usher in the age of the Singularity and the true start of the Cyber Revolution that this book is all about.

One of the key new capabilities that this Cyber Revolution could potentially deliver to the global population is greatly expanded and capable educational and healthcare systems. These services, for the first time, could be delivered to the entire human population across the planet. The engineers think that they have the key to making such dramatic new capabilities available to humanity.

Colleagues at "Geeks without Frontiers" such as Michael Potter, Chris Stott, David Hartshorn and Dr. Delbert Smith have realized that the challenges are not

---

[13] Drew Harwell, "Google's AI assistant sounds human on the phone. Is a warning needed?" *Washington Post*, May 10, 2018 P. A15.

only technological but also involve legal, regulatory, financial, intellectual property rights, and many vested interests that have reason to oppose such a transformation of the world. These are serious issues and ones that mine and Dr. Ram Jakhu have tried to consider in our study that we conducted from 2015 to 2017 with some 80 contributors from around the world. These findings were presented in a book enti-tled *Global Space Governance: An International Study*. The issues considered and possible changes to international law, regulations and so-called soft law were also shared with the U. N. Committee on the Peaceful Uses of Outer Space (COPUOS) in June of 2018 in Vienna, Austria. But these considerations only addressed issues related to satellite usage and the coming expanded usage of the area between com-mercial air space and outer space that we call protospace. There is a need for a simi-lar international undertaking related to the usage of computer technology, artificial intelligence and cyberspace.

The focus of this chapter has been on the trends and possibilities for using global broadband Internet access, AI systems for enhanced educational and healthcare ser-vices and global satellite constellations to expand such offerings around the globe. It has addressed some of the newly emerging opportunities and capabilities and potential problems and issues that will likely emerge as these capabilities expand and improve. It should be clear that these new type offerings are only the vanguard of change. Smart bot-based tele-services that are globally available will rock the world in every dimension. Proto-brain capabilities and services linked to universal broadband connectivity will intrude into every aspect of human life – finance, mar-keting, banking, entertainment, smart housing – and have a huge if not devastating impact on jobs and employment. The world as we know it will change forever. Pay attention.

Chapter 9
Political and Economic Reform in the Fourth Wave

TPeople worry that social media can be manipulated by foreign governments, poisoning democracy and tilting the outcomes of elections. They fear that software algorithms are fueling disinformation, censorship and hate speech. And they are concerned that tech giants have become powerful gatekeepers.

Elizabeth Dwoskin writing for The Washington Post, Dec. 2017

The value of a communications network is proportional to the square of the number of connected users of the system (N^2).

Metcalfe's Law

Advocates of free market capitalism often invoke it as if it were a sanctified and immutable religious belief system that could not and should not ever change. This is nonsense. Humanoids have been around for at least four million years. The smart apeman named "Ardy" (carbon-dated to be over four million years old) and "Lucy" (likewise authenticated to be some 3.5 million years old) now strongly attest to the fact that humanoids have been around for a long time and have been successful in living in social groupings for a long time indeed.

Trade and barter among human social economic systems predated modern capitalism and successfully so for millions of years. Market-based capitalism, based on stock issues and profit-making goals for investors, in contrast, has existed in an organized form for perhaps a hundred years. Further, capitalism as practiced today in the various countries of the United States, Canada, France, Germany, Spain, Italy, Norway, Japan, China, India, Brazil, Mexico, Nigeria, Egypt, Saudi Arabia, Iran and Russia, are all quite different in concept and in practice. They all claim to be societies based on a "rule of law," where products and services are bought and sold in national and global markets, but in many ways these countries are poles apart.

All of these countries have in common individual initiative and individual economic reward for one's efforts. In Norway, perhaps the wealthiest nation in the world on a per capita basis, the differential in the compensation for a corporate

© Springer Nature Switzerland AG 2019
J. N. Pelton, *Preparing for the Next Cyber Revolution*,
https://doi.org/10.1007/978-3-030-02137-5_9

executive and a janitor is typically at most 20 to 1. In contrast, the gap in executive compensation between the top and the bottom in the United States can be well over 1000 to 1. In France, the U.K., Canada and a number of other countries social care systems ensure that *all* people have access to health care regardless of wealth.

As noted in the previous chapter, countries such as France, the U.K. and Canada, are able to accomplish this important social feat of universal health services while also spending *less* of their gross domestic product on health care, on a per capita basis, than is spent in the United States. Although the United States spends over 17% of its federal budget on health care, this huge expenditure actually covers only 45% of the total health care bill. France, the U.K. and Canada spend about 15% of their federal budget on health care, but this covers virtually all of their health care costs. To top this off, the people in France and Canada also live longer. And, infant mortality in many of the OECD countries – is less than in the United States.[1]

Clearly the wonderful, magical and "can-do-no-wrong" marketplace that, in theory, establishes fair pricing that is thought to find a balance between supply and demand is not producing optimum economic results.

We also see a failing of political systems in the age of social media to be able to police against false and distorted news distributed via social media. We have seen the weaponization of information obtained from social media sites to affect the outcome of elections. We have seen organized attempts to create 'fake news' generated and distributed via 'bots' serving as trolls on social media. We have seen the bots masquerading as real people, trying to influence voting in the United States and Europe. This pattern of hostile countries and interest groups engaging in this sort of political activity via social media has been increasing in scope and intensity since at least 2015. The manipulation of social media and 'news' distributed via social media is a true concern. It comes down to a basic question about how to sustain a democratic voting process, and how to avoid large-scale and organized manipulation of the news and the creation of true fake or distorted news in the age of the Internet.

One report from Buzzfeed, an online political reporting network, among many others reporting in a similar vein, explains the huge distortion that a small band of 'trolls' operating via bots to distort their impact can have in generating fake news on social media sites. In November 2016, Buzzfeed reported extensively on the effect of social media on the U. S. presidential elections. Their reports stated that "top fake news reports online had significantly larger impact in terms of measured face time over actual legitimate factual news during the Presidential election." Buzzfeed reported that "engagement on Facebook with so-called 'fake news' as purveyed by bots and trolls, significantly exceeded in hits the combined number of hits for 19 legitimate news outlets."[2]

---

[1] OEDC health care data, 2010, http://oecd.org/dataoecd/

[2] Zeynei Tufekci, "The [divisive, corrosive, democracy poisoning] Age of Free Speech" *Wired Magazine*, Feb, 2018, http://www.wired.com/story/free-speech-issue-tech-turmoil-new-censorship/

Such targeted and weaponized information, distributed via Twitter, Facebook, Instagram and other social media, clearly was effective in manipulating public opinion using false or distorted information. Further, it was disproportionate in its impact. In 2018 a total of 19 individuals from Russia were charged by Special Counsel Robert Mueller as having played a key role in generating a significant amount of this politically charged and distorted information aided by an undisclosed number of bots serving as conduits to reach as many as 50 million Facebook users whose identities and 'political profiles' were disclosed to Cambridge Analytica. This type of asymmetric political warfare, where 19 professional trolls can generate more hits among millions of users than so-called legitimate online news sites, is worrisome indeed. The payoffs were high. Any penalties were modest to non-existent.

This forces the key question: What is wrong with this picture? One theory is that the U. S. form of capitalism has maximized wealth accumulation as a higher goal over all other goals. Wealth is more important than the national health of its citizens and more important than environmental protection or even survival of the species. Clearly wealth is a desirable and a laudable goal, but it is not the ultimate goal of existence.

Among some capitalists, wealth accumulation is currently valued higher than even the survival of the human race. The most advanced economic and political systems around the world have a lot of glitz, glamour and sex appeal. Yet these "advanced" systems, which tend to pursue wealth accumulation as the ultimate goal, are ill-attuned to twenty-first and twenty-second century social, economic and environmental goals. Ask an industrialist to shut down his coal-fired electrical energy plants and replace them with clean energy and the response comes back: "Oh, that is way too expensive." Ask them: "Your money or your life?" The answer seems to come back: "Gee, I can't afford to lose my money. What are the options again?"

Totally free-market systems do not currently include in their product pricing the cost of pollution and wasteful energy usage. Industrial plants and systems that simply do not respond to the needs of Planet Earth are in need of reform. In a recent BBC Sunday morning show, the new U.K. Minister for Energy and the Environment was explaining the need for green, carbon-free energy, but the interviewer was having none of it. She was into pushing her agenda for nuclear and conventional energy plants, over renewables.[3] She kept saying that renewables were much more expensive. What she had not a clue about was the huge accumulating cost of removing carbon from the environment. If we were to double the cost of gasoline and invest it in greenhouse gas reduction it would still not cover the cost of removing the pollution it causes to the environment. This is all part of a paradigm shift that humans will need to make to adjust to living in the second half of the twenty-first century. More about this later in the chapter.

---

[3] BBC Channel 1, Sunday Morning Show, August 1, 2010.

For literally millions of years there were about 100 million humans of several types on the planet. During this time much of what people did was irrelevant to the natural environmental, to natural life cycles or to the sustainability of the planet. Humans were very much a part of a sustainable development ecosystem on Earth for a goodly part of an eon.

When the human population was 800 million – as of 1800 – human industry and activity was still largely irrelevant to the health and sustainability of our planet. When human population hit 1.8 billion at the start of the twentieth century, things began to change. When the population soared to nearly seven billion as of the start of the twenty-first century, human activity began to define the nature of climate change on the third rock from the Sun. We have discovered since the time of Galileo and Columbus that we humans live on a naturally formed spaceship that orbits the Sun once a year. The atmosphere on which we depend is actually quite thin in a relative sense. If we were to view Earth as being an apple, the breathable atmosphere that we pollute more each year is the equivalent in size to the rind of apple-sized planet.

A typical American person, during an average 70-year lifetime, generates perhaps a thousand tons of carbon-based greenhouse gases. By 2100 we may be as many as 12 billion people. This would mean that we humans might all together produce as much as 12 trillion tons of deadly gases. The bottom line is that we humans have, quite simply, become a serious threat to sustained life on Earth, not only for people but for all forms of biota.

Today, humans and human enterprise threaten Earth's natural bio-systems at almost every turn. The list starts with environmental concerns such as noxious greenhouse gases, oil-darkened icecaps, denuded forests, acid rain, polluted oceans and wetlands, expanding holes in the ozone layer and genetically damaging radiation levels. Today, monitoring satellite networks are now measuring some 42 variables related to the environmental health of our planet. All of these measurable ecological factors are moving in a negative direction. Virtually all environmental trends of these measurable variables are negative.

The concerns do not stop there. We are looking at a number of other problems, such as a significant portion of the global population without potable water. This is perhaps the greatest immediate environmental threat to human survival. Already some 25 million people from the Sahel and other arid and semi-arid locations have been forced to relocate to survive due to the onset of droughts accelerated by a growing climate catastrophe. The 25 million people of greater Mexico City are at risk because the natural aquifers under this giant urban sprawl are depleted of water and portions of the land are now collapsing.

A lack of water is only one of existing and coming human woes. Over a billion people now lack access to drinkable water, electrical power or have inadequate access to education and health care. More and more people will also begin to feel the adverse affects of various types of natural resource depletion and unemployment problems.

Some people rail against those who suggest that melting icebergs and shrinking water supplies are real problems. They attack anyone that suggests that climate change is indeed a true danger of our times. These false prophets, backed by often unscrupulous business interests, contend climate change is a 'socialist plot' backed by misguided liberals. They dismiss warnings against global warming and environmental pollution as just being the false warnings of 'Chicken Little' alarmists. Some garner million-dollar incomes by denying climate change and providing disturbing social commentary on our times. They even make fortunes extolling the virtues of gold as a hedge against the 'socialist policies' of Democratic social reformers. These same commentators, however, often forget to tell those that they are advising to buy gold that they are heavily invested in selling gold. There is today certainly a large cadre of economic prophets pushing the purchase of gold. Some these actually have serious financial and investment credentials and, in fact, argue that the only way to survive today's perilous economic times is to purchase precious metals.[4]

So what do all of these problems of environmental pollution, lack of health care and education, lack of water, unemployment and natural resource depletion have to do with capitalism and restructuring of our economic systems to meeting challenges of the future?

Proactive planning and economic restructuring is needed to keep our planet livable, ensure future prosperity and allow our great grandchildren to have a future. We are trying to save the descendents of all people on our planet. We are even trying to save those who would deny that their actions are subjecting all peoples and perhaps all intelligent life forms to extinction.

Modern Capitalism: Its Strengths and Weaknesses

There is a book by Matt Ridley about how modern economic and political systems have always found a way to adapt to social, political and economic needs of the future. Science journalist Ridley's *The Rational Optimist: How Prosperity Evolves* argues that human ideas are essentially like sexual beings. He sees ideas as being subject to natural selection in terms of the best ideas surviving while the poorest and most destructive ideas die a natural death.[5]

In Ridley's world, the best ecological practices and the most sustainable ideas will eventually win out, and, as a consequence, the twenty-second century will find people with a higher standard of living, an ecologically sustainable world with smarter people and improved practices in all areas from transportation to energy to peaceful living patterns. It is a nice dream, and it is a future that sensible and sane people would hope to come true. It is indeed the paradigm shift we need Earth as we know it is to survive.

---

[4] Shayne McGuire, *Buy Gold Now* (2008) John Wiley and Sons, Hoboken, N.J.

[5] Mitt Ridley, *The Rational Optimist: How Prosperity Evolves*, (2010) Harper, New York, New York.

The problem is that we have all seen practices and results to the contrary. We humans have experienced Nazi Germany, the Stalinist Soviet Union, genocide in places such as Rwanda, the Congo and Somalia, and the horrors of nuclear war. We have also witnessed the Great Depression of the 1930s and the Great Recession that took the world economy a decade from which to recover. Both of these devastating economic events were triggered by unbridled greed exhibited by just a small portion of the population. Only a few people, who were motivated by greed and largely coming from the banking and investment world, created world chaos. These events from history suggests that one must perhaps give rationality a little push to realize that the nirvana that Ridley suggests will eventually evolve within the marketplace of ideas. Too often people, given a choice between a good idea and personal greed, will opt for their own wealth. This, too, will become a part of the coming paradigm shift.

In the great competition for the best ideas of the twenty-first century there are some key additional factors to consider beyond the personal greed. One of the important aspects to note is the accelerating rate of technological innovation. In the great intellectual genetic competition of human ideation, we not only see good ideas coming forward faster and faster but also bad ideas thriving and striving as well – often at breakneck speed. Good ideas and scientific knowledge may evolve ever more rapidly but they do not necessarily take hold quickly among the public. In the United States a staggering number of people do not believe in evolution some 200 years after Darwin's careful research had been completely verified. Apparently somewhere around 40% of Republicans in the United States do not believe in evolution. There are millions of people that, for instance, do not accept the reality of climate change, who think the Moon landing was faked, or that the creation of the universe and Earth was a 7-day event, exactly as recorded in Genesis in the King James version of the Bible.

Twenty-first century thoughts can not only move us forward but backwards as well. Bigotry, racism, creationism, anti-intellectualism and general stupidity is certainly alive. Competition for the best ideas in the global ideation marketplace remains a free-for-all.

We certainly know that innovative thought can create very bad ideas, such as weapons of mass destruction, pandemics and environmental disasters. Change is happening at exponentially increasing rates, and this means the consequences of competing ideas – both good and bad – impact the human marketplace with ever greater speed with less and less tolerance for error.

Further, not only are new ideas being conceived but technology systems are aiding their implementation within a global society. Global information networks, high speed transportation systems and intellectual property rights authorizations are allowing for these ideas and products to be unleashed on the world at incredible speeds and within every nation.

In the past, a bad idea or trend could founder in geographical isolation. The Black Plague was devastating, but it still ran its course in a relatively limited geographic area. Today, a global pandemic can spread with incredible speed via airline travel.

The phenomenon of future compression, instant global impact and the volume of human enterprise and its impact on the global biosphere are all unprecedented in the history of humankind. Mistakes in the past could be overcome by a slow rate of geographic spread, ineffective technological development, or the opportunity to die out over time. We no longer have these inherent protections against bad ideas. Further unfettered capitalist principles – driven by excessive greed and instantaneous electronic marketing – can amplify the dangers of human activity.

The dangers of an overpopulated world, tied together by communications and transportation systems with huge populations concentrated along seacoasts, should be abundantly clear. The twenty-first century world will of necessity live in growing fear of global pandemic diseases, of attacks by biochemical or nuclear weapons, of killer earthquakes, and devastating weather events such as typhoons, hurricanes, tornados and tsunamis. One might say, humans have always had to face natural disasters, but climate change can intensify energy in weather systems. Further, as populations expand and urbanization grows to include the majority of all people on Earth, vulnerability also increases. We have yet to learn the nature of the many unintended ecological and environmental disasters that could come from the polluting industrial effects of an overpopulated world that is eating up natural resources at a prodigious pace. In short, we currently face a wide range of megacrunch challenges that range from ecological disaster and super urbanization to technological unemployment to unchecked automation of all human processes. Without some checks and reforms to unrestricted capitalist markets we humans may well have a tough time surviving even the rest of the twenty-first century (Fig. 9.1).[6]

Motivations for Re-inventing and Restructuring Capitalism

The bulk of European society has, at least to some degree, embraced the concept of sustainability as a goal to work towards over time. Over 50% of Germany's electrical energy now comes from renewable energy such as wind, solar, hydroelectric, tidal, etc. The city of Copenhagen has moved since the 1970s from greenhouse gas emissions of about 15 tons per person down to 3 tons per capita. Furthermore, they have specific objectives to become a zero carbon emissions community.

To understand the success of Copenhagen's journey toward becoming a sustainable city one must explore initial motivations. The motivation for change in Copenhagen came during the gasoline crisis of the late 1970s when very limited supplies drove the cost of gasoline to historic highs, and, in some cases, there were simply not enough supplies despite rationing. The leadership responded to the crisis

---

[6] Joseph N. Pelton and Peter Marshall, *Megacrunch: Ten Survival Strategies for the twenty-first Century* (2010) PA Associates, London, UK.

Fig. 9.1 Mega-crunch
challenges of the
twenty-first century and
capitalist reform

by developing strategic plans to both reduce energy demand and to develop alternative and renewable energy supplies. The decision to 'go green' was really not about sustainability. The basic motivations were economic and strategic. The Danes vowed to become free of dependence on foreign oil and find energy efficiencies that would over the longer term provide significant economic savings.

This is really the type of good idea that the capitalist marketplace needs to be fed as it adjusts to twenty-first century needs. This is a truly a 'win, win, win' type of idea. Today, the people of Copenhagen have buildings and transportation systems that use less energy. Co-generation makes energy systems more efficient. Renewable energy gives the city and the country strategic independence from foreign oil sources, and after 35 years of transition significant savings are being achieved. The positive environmental effects are almost an additional dividend. The bottom line is that Copenhagen's political leadership decisions have produced at least three wins. The decision to cut energy use and pursue clean renewal energy systems for housing, buildings, transportation, etc., has: (a) produced major economic savings; (b) provided strategic independence from foreign energy supplies; and (c) helped to clean the environment.

This motivation to move towards clean sustainable energy practice did not require a restructure of capitalist practices. It only took a far-sighted act of political will. Today the action by German political leadership to move strongly toward clean renewal energy is more controversial because savings have yet to be achieved, and there is currently no major foreign oil supply crisis. The bottom line is that the consuming public does not think and act strategically. Unless there is a strong economic incentive, people are more likely to follow the cheapest short-term route.

The difference in European capitalist practices and the U. S. capitalist market-place is clearly revealed with regard to automobiles and gas taxes. West European countries, for at least 20 years, have imposed high gasoline taxes – at least equal to the cost of the gasoline and sometimes even more. The revenues have been used to support mass transit and other such projects. The result is that the fleet of cars on the road in Europe is generally small and very fuel efficient. In contrast to Europe, the United States has very low gas taxes, and the cars are generally larger and much less fuel efficient. U. S. President Gerald Ford signed into law a requirement for U. S. car manufacturers to raise the fuel efficiency of their fleet. It was some 35 years later that President Obama managed to raise U. S. gasoline mileage requirements to more stringent levels. Ads are run by oil and gas interests to tell people that higher U. S. gasoline taxes will mean that they will suffer.

The problem with the U. S. capitalist system is that too often large business interests intervene with slanted messages that often mislead or misinform the general public about strategic opportunities and their longer interests to transition to new systems that can make their lives better over the longer term. Big business, particularly in the United States, by focusing on short-term quarterly profits, often shoot themselves in the foot – and the U. S. public as well. Solar, wind, geothermal, fuel cell, advanced battery systems and hydrogen-based energy systems over the longer term can and will save consumer money, clean the environment and allow strategic independence from foreign energy sources. Shorter-term lower energy costs, industry misinformation and political interaction stand in the way of the longer term best public interest.

The key is finding a way to provide economic incentives to get people to do the right thing – based on their longer-term interest. One option would be an environmental tax that could be returned to consumers who invest in renewable energy, high mileage cars, or well insulated homes and businesses. Forty percent of energy in the United States goes to heating and cooling homes and buildings, and 25% goes toward transportation. Clearly steps are needed, within the arena of good ideas, to get people to cut the energy expended on buildings and transportation and acquire new sustainable energy systems. Dedicated taxes that spur more efficient energy and water usage, encourage recycling and discourage the use of oil and gasoline could be structured so that people who go green would not only get their tax investments back but actually save money in the longer run. These economic incentives, however, will not cover all elements of environmental reform. Building standards can certainly be set to save energy and cut costs. Throughout Europe, most building lights are activated by motion detectors and escalators are called into action by a tread plate. Although such sensors cost money to install they pay back the investment many times over during the lifetime of a building. Energy-conscious insulation codes, emission inspections on automobiles, trucks and buses clearly pay off for the public and individuals as well.

The bottom line is that economic incentives coupled with building and energy codes clearly work. Political leaders must be willing to encourage behavior that will save consumers money and allow them to be more secure over the longer term. Dedicated environmental taxes that are rebated to consumers are one way to restructure capitalism without a sea change in capitalist economic principles. The first step in this transition clearly needs to focus on energy and transportation because of the urgency of addressing climate change dangers to the human race and the solid economic returns that can come from renewable energy systems and energy-efficient buildings and transportation. There are other second-tier issues that almost inevitably follow. One of the key aspects that looms on the horizon is that of technological unemployment.

Capitalism, Technological Unemployment and Modern Cultural Clashes

Ironically, while Europe and other OECD countries such as Japan lead the way in adapting modern capitalist systems to energy efficiency, these same countries may not be well geared to coping with technological unemployment. In the case of super-automation current economic incentives can often backfire. More and more technological innovations, if not channeled in productive ways, can serve to exacerbate the problem of full employment and societal prosperity rather than solve it.

European capitalism today is almost a nightmare for companies seeking to be economically competitive on a global market. In France, for instance, retirement age has fluctuated between 50 and 55. Working hours per week have stayed around 35 h. Employment is for life. An employee let go is eligible for retirement pay for the rest of his or her life. Further, strikes among workers can come at any time. Not only are there scheduled strikes but wildcat strikes or sympathy strikes can occur like a bolt of lightning. What is the result of these considerable employment issues and problems? One clearcut result is to eliminate jobs.

When French firms seek to compete against firms from Taiwan or South Korea, where workers typically work from 52 to 54 h a week, the challenge is clear. In a French restaurant your waiter is your cashier. Your bill is presented along with a handheld electronic device in which you can swipe your card and receive a receipt. At a French Auchan superstore, you can buy everything from a high-definition television to a bag of potatoes to a condom. As you enter the store you swipe your card through a machine, and you acquire a handheld device that you scan into your electronic shopping card for everything you want to buy. When you leave you simply go to the rapid checkout and you are done in a flash.

And then again there is Amazon and Ali Baba, where all the purchasing is done online without a retail clerk in sight. In fact, at the new Amazon stores, you purchase a product by taking it from the shelf and putting it into your electronic shopping cart.

Again, retail clerks are disappearing left and right. In many socialist-leaning countries, where people work short weeks, retire early and essentially are impossible to fire, jobs are shrinking as super-automation surges forward. Occupations from accounting to pharmacy, from secretary/receptionist to property appraisers, are going the way of telephone operators and stagecoach drivers.

The theory of capitalist economies is that as technology and innovation closes one door another one opens. The trouble is that new jobs are often not matched with the skill, qualifications and compensation of jobs of the past. Some of the most prevalent new jobs of the past 20 years have been teaching aides and nursing aides, but these professions often pay less than jobs of the past such as auto assembly line workers or pharmacists. Some analysts suggest that the trend is for fewer and fewer highly skilled professions and more and more unskilled jobs that pay very poorly. If this trend should hold, the future would be dire indeed. In his first book *Player Piano* in 1952 novelist Kurt Vonnegut, Jr., projects a dystopian view of the future. In this future society, the role of the mass proletariat is simply to consume a never-ending stream of poor quality consumer products with very little meaning to their lives. The rapid onset of super-automation could very well portend problems of technological unemployment, dead end jobs and a highly stratified society of haves and have-nots.[7]

The problem is that one of the prime objectives of capitalist economies is to produce technology that does more with less and thus reduces the need for human labor and accordingly eliminates jobs. This capitalist use of technology increases throughput, productivity and profits, which sounds all to the good. Unfortunately, the elimination of jobs or the "de-skilling" of human labor can also eliminate consumers able to pay for products or services that the more "efficient" economy is producing.

The great dilemma of twenty-first century capitalism is that productivity gains can also undercut a vibrant consumer base and potentially increase the ranks of the unemployed or underemployed. This technological unemployment is not only bad for the economy but can, if sustained, lead to political instability as well.

Another key twenty-first century problem is that the whole Western concept of industrialization, technological and scientific advancement, productivity gains and ever- increasing rates of throughput is not accepted as the norm – or even as a reasonable societal objective – by a large portion of the world population. Traditional religious leaders often contend that Western economic and social practices break down religious beliefs and familial and cultural relationships. They oppose not only many aspects of technological advancement and modernization but what are often

---

[7] Kurt Vonnegut, Jr., *Player Piano* (1952) Charles Schribner and Sons, New York.

seen as modern social values such as equality of the sexes, gay rights, universal education, separation of church and state, and democratic political processes.[8]

Author Benjamin Barber describes the clash between capitalism and the world of tribalism, the thrust of 'modernism,' as having great force behind it. But he also questions whether what he calls McWorld will naturally sustain democracy when confronted with such threats as unemployment, ethnic rivalries, tribalism, or fundamentalist religious calls to revolt. Barber indeed suggests that "savage capitalism" remains vulnerable to anti-democratic revolution when consumers are offered the choice of "nine types of VCRs" but their salaries "are not sufficient to pay for the rising cost of bread." He concludes, however, that. "... McWorld's homogenization is likely to establish a macropeace that favors the triumph of commerce and its markets and to give those who control information, communication and entertainment the ultimate (if inadvertent) control over human destiny."[9]

In reading books of philosophy and political history, it becomes clear that it is important to distinguish between higher social and cultural goals in contrast to mere economic systems of trade and commerce. Higher level goals such as survival of the species, personal security, plus liberty, equity and fairness, justice, individual privacy and freedom are key ideals that are to be aspired toward and hopefully achieved within a free society. Systems of commerce and consumerism are clearly important, but they are merely that. A society with well-stocked shopping malls and supermarkets are nice to have, but ease of shopping or low consumer prices are not worth sacrificing liberty or fairness in society.

If half of a society is wealthy and prosperous and the other half is poor, needy, out of work or engaged in violent and criminal behavior, the chances of key social goals and individual liberties being achieved are very small. Efficient capitalist systems can even serve as a threat to democratic values. As Barber has observed, malleable consumers who respond to competitive advertising help maximize economic throughputs in capitalist economies. Such people probably do not make the best independent thinkers and voters. Those capitalist consumers who are committed to a 'shop until you drop' world are not the most likely candidates who will fight to protect civil liberties or serve as prime advocates for green and sustainable communities free from pollution. Most corporate executives are not tree huggers. Most dedicated greens or members of the American Civil Liberties Union (ACLU) are not out shopping for $2500 Gucci Sshoes or sumptuous Dior ball gowns.

The future of capitalism is thus in some ways more perilous than at any time in recent history. The clash between prime democratic values and consumer-driven capitalism continues to heat up for the following reasons: (a) climate change, global warming, ozone layer depletion and sustainability issues arising from global natural

---

[8] Benjamin Barber, *Jihad vs. McWorld: How Globalism and Tribalism Are Reshaping the World* (1995) Ballantine Books, New York. A similar type of analysis of the rift between traditional and modern cultures is provided in Thomas Friedman, *The Lexus and the Olive Tree (1999)*, Farrar, Straus, Giroux, New York.

[9] *Ibid.*

resource depletion and pollution; (b) automation and technologically driven productivity gains that serve to fuel technological unemployment, under employment and 'de-skilling' of the broader labor force; and (c) unchecked swings in global recessions driven by excess speculation by investment bankers, derivative traders and real estate speculators.

The advocates of unbridled *laissez faire* capitalism argue that open competition and ongoing productivity gains will always produce the best of all possible worlds with more prosperity for all. Commercial nabobs with a commercial zeal that even Ayn Rand would find laudable contend that throughput and technological innovation will always produce a better world and that anyone that argues otherwise must be a grenade-tossing, Bible-hating Communist revolutionary. Such zealotry tends to overlook one important thing – reality.

The free market at all costs crusaders blithely overlook such details as global warming, holes in the ozone layer, drought in the Sahel and other major deserts, destruction of the world's rain forests, rampant teenage pregnancy, spreading global drug addition, an AIDS epidemic that has now spread to rural China and 50-year highs in unemployment for many OECD countries, as well as an absence of water, electrical power and adequate healthcare or education for billions of people around the world. This is the form of capitalism that says: "My refrigerator is full and my swimming pool filter and air conditioner is working, so the world must be okay." Certainly today's problems have in a way snuck up on the affluent societies over the past century.

Only a hundred years ago – a mere pittance of time as measured against the four million years of human history – the world was still relatively pristine, and our bio-systems reasonably sustainable. Wildlife and plant life flourished. The ozone layer protected all fauna and flora against radiation. Some two billion people and the related industries and activities of humanity had but a modest impact on the world's biosphere. We committed many indiscretions, but fortunately Earth was capable of handling it with relative ease. The only real threat to human survival as of 1900 was correctly seen as being external to our world – such as a comet or asteroid colliding with Earth or an alien invasion. It has taken but one century to change all that. Another heedless century of overpopulating Earth with people and unchecked industrial practices may be sufficient to do us all in – permanently.

How do we change? We change our economic systems. We create economic incentives for people to do the right thing. We teach our young the cost of unchecked greenhouse gases. We create "carrots" that provide incentives for green transportation, green buildings, and green energy systems. We have also started to use a few "sticks" to discourage pollution of the air, water and oceans, the wasting of energy and the burning of coal, oil and wood. In short, we have begun to minimize overconsumption in our twenty-first century world, but this is just the start and not the finish. Author, social-economist and essayist Tom Friedman tells us rather bluntly, but rather convincingly, that the twenty-second century world we inherit will be "hot, flat and crowded"[10] (See Fig. 9.2).

---

[10]Thomas Friedman, *Hot, Flat and Crowded* (2008).

Fig. 9.2 Soon more than our beaches will be overcrowded

One of the keys to the future is of course to make green industries profitable and green and renewable energy cheaper than dirty industries. It is here that technology can shine. Part of the key is changing the time perspective of consumers who have been conditioned by capitalist industry to think and act in the short term. Transportation and energy systems plus housing and offices need to be designed and built for the longer term. If such systems were designed not to maximize consumerism but to minimize cost and maximize durability and recycling efficiency, huge savings could be achieved and the world would be less polluted and more livable.

An Agenda of Action

Many people at this stage are likely to ask: So what are the specific steps we need to take? Few people are ready to buy into any reform until they know: (a) What is being proposed; (b) How much it will cost; (c) What the pros and cons are; and (d) Whether or not it will really work and how long it will take. The following is a sample list of steps that could be taken now to start reforming totally market driven capitalism (See Table 9.1).

The above concepts and ideas are simply that. Table 9.1 represents only a preliminary listing of future directions that we need to work toward if we are to survive as a species and if we are to use the benefits of super-automation and new technological innovation to advantage.

Table 9.1 Steps forward to a more flexible and responsible global economy

A new agenda for restructured and reformed twenty-first century capitalism		
Proposed action	Implementation strategy	Time table
Tax and financial incentives for homes and offices of the future	**Activity.** Scale projects of 5000–10,000 housing units or similar number of offices in building complexes. Energy use of less than 30,000 BTUs a year per cu meter of space. Latest in renewable energy and co-generation, insulation, recycled building materials, trash removal, water renewal/gray water use, lighting fixtures, etc. Private builders to compete for contracts to build these complexes. Can be redevelopment or new developments. Must include parks, retail, schools, and transport geared to latest environmentally friendly technologies, pedestrian walkways and canals. **Implementation incentives.** Tax credits or deductions. 20% subsidies toward rent or purchase for first 5 years of occupancy with a 20-year commitment (that can be resold to government on a sliding scale) **Financing.** Public/private partnerships. Billionaires and foundations to be solicited to help with capital financing, especially for schools and parks. At least one project for every 500,000–1 million population worldwide	**2020–2035**
Migration of technology of houses and offices to all structures via revised building standards and incentives. Standards would sunset every 10 years with mandatory updates	**Activity.** Implement federal building standards where possible. Provincial or local standards when necessary. There would be standards based on "House and Offices of the Future" Technology. These would be comprehensive and phased in over the next 10 years to be mandatory for all new construction. Fifteen-year mandatory retrofit for existing structures. Standards would cover insulation, lighting, heating, ventilation & AC, use of recycled materials for building, roofing, water, sewage and trash, windows and skylights, fireplaces, renewable energy systems, and major utilities (washing machines, ovens, refrigerators, etc.) Emphasis would be on long-term efficiencies **Implementation strategies.** Complete payback of investment within 20 years and design for at least 50-year life cycles. Tax credits and/or deductions for early conversions. Conversions can be paid for in an amount equal to 75% through green tax rebates (see below)	**2020s–2030s**

(continued)

Table 9.1 (continued)

A new agenda for restructured and reformed twenty-first century capitalism		
Proposed action	Implementation strategy	Time table
Transport incentives and taxes	**Activity.** Since about 25% of energy consumption and greenhouse gas pollution comes from transportation, there need to be incentives to use clean mass transport or vehicles and strong disincentives for using dirty vehicles. This policy would also strongly encourage people to live near their work or to telecommute **Implementation strategies.** Subsidize all mass transit at $1 a ride. Financing for this subsidy would be through gasoline and car ownership taxes. First car per household would be 4% of value per year (unless this car met green car standard of 50 km/l of gas consumption, then rate would be 2%); second car would be 8% of value per year (unless it was designated a green car and then the rate would be 4%). The third car would be 12% of value and so on. There would also be a federal tax on airline trips that would be geared to the aircraft in terms of size, kilometers travel and how green or dirty the airliner actually was. Green buses and trains would be exempt from this federal tax. There also might be provincial or local taxes, but it would be useful to have federal taxes that were consistent around the world	**2020s–2030s**
Legislation to protect personal privacy and prevent foreign interference in elections	In Europe the General Data Protection Regulations have now entered into effect. In the United States efforts to create a Consumer Privacy Bill of Rights have been defeated by Internet-based industries. New legislation is needed to protect against foreign interference in elections through the use of social media as well as new protections against the collection and use of information by national states and social media and Internet-based retailers	**Now–2030s**

(continued)

Table 9.1 (continued)

A new agenda for restructured and reformed twenty-first century capitalism		
Proposed action	Implementation strategy	Time table
Green tax	**Activity.** Governments impose taxes essentially as a means of raising revenues to operate. There is an increased realization that tax policy and/or standards are a means that governments have to influence social and economic behavior. There are ways that government can use tax policy to pursue socially and environmentally desirable behavior and pay for the cost of pollution clean up by dedicating a green tax to perform actions for the common good **Implementation strategies.** Under this type of green tax 90% or more of the collected revenue could be reinvested in things such as the "House or Office of the Future" building standards. The beauty of this type of tax is that it could allow the citizen taxpayer to constantly save money on electricity, heating, air conditioning, etc., and if a certain standard of efficiency were ultimately met, they could opt out of paying this type of tax altogether. Indeed at this stage they might even be selling energy to the grid and thus earning income, and the value of their upgraded green home might have escalated to the point that they could sell their property at a handsome profit	**2020s**
Eco tax	**Activity.** When consumers buy products at the grocery store they can see the ingredients, the calories, the volume and the price. What they do not see is that 500 g of steak or frozen foods might cause 50 times more net pollution than 500 g of vegetables. They also would have no easy way to tell the environmental impact of buying a gas lawn mover versus an electric mower. The cost to the public of removing pollution from the atmosphere or the water is increasingly very real. Buying 500 g of steak is like driving 100 km in a car in terms of atmospheric pollution. The answer to this problem would be an eco tax **Implementation strategies**. With RFID technology and computers various products could attract an additional sales tax equivalent to their adverse environmental impact. Products such as steak, frozen foods, paper, lumber and coal would suddenly become far more expensive. Other products such as fish, chicken, vegetables, bricks, etc., would be taxed at a very low rate	**2020s**

(continued)

Table 9.1 (continued)

A new agenda for restructured and reformed twenty-first century capitalism		
Proposed action	Implementation strategy	Time table
Phasing out coal-fired electrical generating plants	**Activity.** Operators of coal-fired plants would be given an economic incentive to phase out coal or petroleum-based electricity-generating plants within 5 years and a bonus if they did it in four. The key would be to replace the facility with a zero-carbon facility **Implementation strategies.** This would be funded around the world by eco or gasoline taxes	**2020s**
Redefining work and value in a global society	**Activity.** For several centuries we have defined work as something one does for wages or some form of capitalist enterprise. In the age of super-automation, where machines can and essentially will do a growing percentage of all work, we need a paradigm shift that restructures our economic systems. If we do not overpopulate the world, machines can do more and more of our work, create value, produce food, undertake manufacturing, as well as carry out services and also innovate and create new technology – faster and perhaps better than humans **Implementing strategies.** People will increasing spend their time in being educated, working with intelligent machines to carry out planning, engaging in the arts of all types, seeking to create off-planet worlds for human society, and creating a peaceful and plentiful world. Gradual population reduction over the next century becomes a key goal. The world that John Kenneth Galbraith envisioned a half century ago in the affluent society starts to potentially become a reality if the paradigm shift and the restructure of capitalism can be achieved. The key is a shift from "value" being defined as individually accumulated wealth to attainment of greater knowledge, climate restoration, Earth-wide affluence and creation of both a sustainable Earth but also sustainable off-world societies	**2020s–2030s**

Will all of these possible solutions be implemented globally and in rapid fashion? No, certainly not. The world of politics works slowly. Action along these lines is more likely to happen sooner in Europe and Canada and perhaps Japan and Korea than in the United States.

Time will tell if ideas of this type can build a new economy with new values. Time will tell if such a paradigm shift might actually build a sustainable, advanced technological society. Time will tell if work and economic value and the actual nature of work can be successfully redefined. Time will tell if life-threatening pollution and population numbers can be reduced so that there is enough water and resources for everyone – people, plants and animals – for us all to survive. Time will tell if it is possible that the United States and other capitalist countries locked into the mentality that a "free market" run by investment bankers is the best way forward in the second half of the twenty-first century.

Conclusions

The existence of various types of economic systems can be thought as a sort of laboratory where we can experiment over the rest of this most crucial century with the nature of work, economic value and societal efficiency. For humans, the hardest part will be thinking of increasingly smart and capable machines in new ways. For at least several centuries, people have seen machines as ways for us to accomplish work and carry out physical labor for us more efficiently. Now we are faced with a new paradigm. We are faced with machines that may be smarter than us and can think faster, more clearly and – most distressing to our egos – more creatively than ourselves. We have been the thinkers and our machines our working assistants. We are really not prepared for this brave new world in which our machines do most of the thinking and innovating and we really do not know how to assign value and even to define work in such a psychologically distressing world of the future. But adapting to such a world is the challenge of the next few decades.

For humanity the future is not only now but increasingly urgent.

Chapter 10
Brave New World

As the most important phenomenon in the universe, intelligence is capable of transcending natural limitations, and of transforming the world in its image. In human hands our intelligence has enabled us to overcome the restrictions of our biological heritage and to change ourselves in the process.

Ray Kurzweil, *How to Create a Mind*

Introduction

To many people any 'future' that extends much beyond next week or next month may tend towards becoming a hyper object. Perhaps you will recall that we talked about hyper-objects in Chap. 1. This is the concept first defined by Tim Morton. It is so vast in scale and impact that people say this is just too complex to think about. Somebody very smart needs to sort this out and find the answers. If one takes the time to read the U. N.'s 17 Sustainable Development Goals for 2030, as presented in Appendix 1 of this book, you might say these concerns really seem just too complicated to understand and figure out in a rational way. The assembled facts in the U. N. reports about population growth, rising pollution levels of the oceans, the worsening condition of the atmosphere and the land, the increasing levels of human consumption of food and energy and the incredible rise in the need for more jobs, more water, more food, and more education and healthcare services pose some very daunting tasks. The U. N. Goals set for 2030 are huge, and how they all interact are truly hard to comprehend.

These U. N. goals seek to alter the future of human society on Planet Earth on a grand scale, and they urge governments and organizations to make big changes in less than two decades. When taken together as a whole they represent a giant dilemma – how do we rationally make such changes and reorient global behavior for world society, especially when many of the goals are contradictory. How can we pursue economic growth, industrialization and more jobs while also seeking environmental reform and combating climate change? How can we preserve Earth

© Springer Nature Switzerland AG 2019
J. N. Pelton, *Preparing for the Next Cyber Revolution*,
https://doi.org/10.1007/978-3-030-02137-5_10

as a viable place for humans to live for the longer term without changing some basic practices? Many of the recipes that were previously baked into the U. N.'s Millennium Development Goals are now reiterated in the Sustainable Development Goals.[1] The future of human existence is, in a word, complicated. If one looks at the United Nations' 17 Sustainable Goals, not one at a time but as a whole, one can see that many of the goals are in conflict. Economic expansion, the creation of more jobs, the growth of human population and the expansion of the modernized offerings to be found in the large cities are either implicitly or explicitly contained in these goals. Those objectives can often be in conflict with the environmental protection goals that apply to over half of the 17 goals. Further, the various goals related to the development of more technology and scientific knowledge could easily lead to more industrial throughput and expansion associated with a 'disposal economy. 'Indeed these modern innovations could also easily lead to the shrinkage of jobs and increasing pollution of the oceans, the atmosphere and inhabited land.

So does this mean the situation is hopeless? No. There are clear and viable solutions, but they require better answers and better economic incentives to convince people to change. In short, we need to face certain difficult truths. And what are those truths?

First, the enemy of a better human future is excessive population growth. We are over exploiting our forests. We are overfishing the seas. We are over mining and consuming Earth's limited resources to meet the needs of a human population that has simply grown too large. Priority must be given to finding better ways to control population growth and the negative aspects of excessive density in super-crowded megacities. We also are in need of creating circular economies that recycle resources rather than continuing disposal economies that simply consumes and trashes valuable resources. Incentives to limit family growth plus encouragements for sustainability, recycling and use of clean energy systems – these are the hallmarks of a Fourth Wave economy that might actually work.

Second, what we need to address in a serious and sustainable way is the issue of climate change. We need to address not only climate change but all of its consequences such as global warming, sea level rise, melting of the Arctic ice, violent storms, desertification and loss of arable lands, etc. The key here is to create a host of new technologies, private enterprise and industries and employment opportunities that make combating climate change not a scourge to endure but a great new opportunity for prosperity, new jobs and industrial expansion that blossoms with a new Fourth Wave economy,

Third, we need to address not only the great potential of technology but its powerful effects on global society. Technology can help us create a future of abundance, develop new proto-brains of great ability, blanket the world with broadband networking systems and bring education, healthcare and a wide range of services to the world. In this regard technological innovation is in many ways the arbiter of the future.

---

[1] United Nations Sustainable Development Goals, Targets and Facts and Figures, https://www. un.org/sustainabledevelopment/development-agenda/

But we must also consider the negative side of the ledger. Technology can also fuel the speed of climate change, increase the rate of unemployment and pollute Earth. Hidden within this powerful technology and the new automated infrastructure that allows it to operate around the globe are also increased vulnerabilities. These include not only vulnerabilities to cyberattacks, but also natural disasters and cosmic hazards such as asteroids, solar flares and especially coronal mass ejections that can fry our electronic power grids, global network of pipelines and the satellite networks on which we depend.

Our society of some 7.5 billion people are blithely unaware of how many deadly dominoes could fall if modern society were to be hit by a blast from the Sun that equaled that of the Carrington event of 1859. The study by Lloyds of London of such a scenario came up with an estimate of a $3 trillion event. However, the results of a cosmic assault on Earth could be far, far worse. This is because the Lloyd's of London study did not take into account the new vulnerabilities that are expected by the shift of the world's magnetic fields. Today's magnetic fields shield us from the solar wind and especially deadly coronal mass ejections (CME's) that endanger our electronic power supplies, or pipeline controls, other types of industrial controls and our satellites.

The arguments presented in this book end with a fervent plea that we have to be more careful. The powerful tools of technology must be deployed more wisely and strategically to address urgent needs.

In short, we need to recognize the importance of re-deploying technology to cope with the most important of twenty-first century challenges. The more thoughtful uses of technology involve a series of choices, difficult choices. It requires us to look at documents such as the unanimously adopted U. N. Millennial Development Goals, and now the Sustainable Development Goals for 2030, and realize that we must face up to the omissions and the contradictions. The omission is the criticality of focusing on family planning and reduction of the global human population as a chief cause of poverty, hunger, unemployment, global warming, climate change, pollution, and resource shortages, such as potable water.

As the Millennium Development Goals were being assessed back in 2006, and noting the lack of success with family planning, the following observations were made:

> *Promotion of family planning in countries with high birth rates has the potential to reduce poverty and hunger and avert 32% of all maternal deaths and nearly 10% of childhood deaths. It would also contribute substantially to women's empowerment, achievement of universal primary schooling and long-term environmental sustainability. In the past 40 years, family-planning programmes have played a major part in raising the prevalence of contraceptive practice from less than 10% to 60% and reducing fertility in developing countries from six to about three births per woman. However, in half the 75 larger low-income and lower-middle income countries (mainly in Africa), contraceptive practice remains low and fertility, population growth and an unmet need for family planning are high. The cost-cutting contribution to the achievement of the Millennium Development Goals makes greater investment in family planning in these countries compelling.*[2]

---

[2] John Cleland, Stan Bernstein, Alex Ezen,mAnibal Faundes, Anna Glasier, Jolene Innis, "Family planning: the unfinished agenda" *Lancet,* Nov. 1, 2006, https://www.thelancet.com/journals/lancet/article/PIIS0140-6736(06)69480-4/fulltext (Last accessed May 15, 2018).

Unfortunately the situation with population growth has only become more acute. As noted with regard to Sustainable Development Goal #11, on sustainable cities, the projection is that 95% of the growth in urban areas will be in economically developing countries. This growth will unfortunately represent areas characterized as slums, with massive levels of unemployment, pollution, and huge challenges with regard to education, healthcare, and with problems such as access to drinking water, sewage, and housing.[3]

The compromises that were reached in September of 2015 that led to the unanimous adoption of this document by 193 nations includes antithetical compromises. Higher and higher rates of industrial throughput, more jobs and decent wages and more technology that is today designed to sell more products will be largely counter to ten of the seventeen goals that touch on environmental preservation and coping with climate change.

Critical Choices to Cope with the Coming Cyber Revolution

It is essential for political and economic leaders to face certain self-evident truths. The uses of technology today are fundamentally geared to support population growth and disposal economies. The objective thus becomes higher and higher rates of throughput, creation of wealth and more and more consumption by an ever-larger population that fuels growth.

This essentially comes down to what space-philosopher and inventor Neil Ruzik said about the human race: "We are the Big Eaters." He noted this on "The Voyage Beyond Apollo" cruise that went down to watch the last Apollo launch in 1972. On this cruise to watch the lift-off of Apollo he gave a thought-provoking talk. He said simply this is what people do. We eat, and eat, and eat. And then we procreate to produce more people so that they can eat and eat some more. He explained that if we go to the Moon and Mars we will eat and eat the resources of these cosmic bodies as well.

Perhaps a new generation of humanity that we might call *Homo electronicus* will someday – and someday soon – recognize that this escalator can take us to our doom. Perhaps starting with the millennials will recognize the need for us humans to evolve into new modes of behavior. We truly need to devise a Fourth Wave economy that is optimized by a high tech/high touch world that is based on circular rotation of resources rather than their consumption. We need to recycle and sustain our world in many ways. This means to convert our energy systems so that they are solar, tidal, hydroelectric, geothermal, or use new systems such as ocean thermal energy conversion (OTEC) to heat and cool our homes and electric cars, trucks, buses, boats, planes and trains to move people and our freight. We also need to invent a cleaner way to get to space or power hypersonic transport through the

---

[3] United Nations Sustainable Development Goals, Goal 11, Facts and Figures https://www.un.org/sustainabledevelopment/development-agenda/

stratosphere. This is because the stratosphere is more than a hundred time easier to pollute than the atmosphere at sea level. We need to stop the burning and use of any hydrocarbon fuel and find a way to avoid putting any greenhouse gas into the atmosphere and find a way to halt gases such as methane escaping from mines, cattle digestion, frozen peat fields, the Arctic regions and oceans and rising into the atmosphere. Regulations are one way to do this, but economic incentives and new technologies or systems that allow us to avoid being polluters of the atmosphere and increasing climate change threats will likely work the best, including finding workable incentives to produce billions less children being born on Planet Earth.

If we could take things down a notch and produce only about 1.5 children per family through the year 2100, this would gradually bring us back to an optimum number of around five billion people. This may seem like a reasonable target, but it is of course hard to produce 0.5 children. It is a rather digital equation of either 1 or 0. The real problem is that in developing countries where agriculture represents 40% of the jobs, families tend to produce larger families to work the fields. Until there is a clear economic rationale and program that can show families that having children is going to keep them poor and there is a better future than subsistence farming, there will be more and more kids produced.

Tax policies can reduce the size of families in economically advanced countries. However, in economically developing countries, programs will probably need to be village-based to work. This means creating incentives and perhaps even penalties (i.e., carrots and sticks) to encourage smaller families. Villages with smaller families might be rewarded with electrification, lighting, schools and clinics and especially village-based jobs. The real question is who at the village, provincial or national level would pay for the incentives and set the goals? National governments, such as those of India and China, which have tried to set a quota for family planning have not exactly been rewarded for their efforts.

The Cyber Revolution and Climate Change

There are still people out there who are climate change deniers. They note that the ice ages have come and gone over millions of years and that the last period of major glaciation occurred as recently as 40,000 to around 11,000 years ago. They note that the reasons for the ice ages and the current modern warming period have not been fully explained.[4]

However, we do now have a fairly sophisticated understanding of how Venus was transformed into a pressurized gas fireball by means of trapped gas that continued to be heated to higher and higher temperatures. This created higher and higher pressures that led to a deadly and finally lethal cycle.

---

[4]"A New Picture of the Last Ice Age", *Science Daily*, March 17, 2016. https://www.sciencedaily.com/releases/2016/03/160317095002.htm

On Earth we now know about the continuing buildup of greenhouse gases. We know that these carbon-based gases, such as CO_2 and methane, are now controlling not only the heating of Earth's atmosphere but how the process is feeding on itself. Thus we see icebergs turn into salt water that will not refreeze. We are seeing how petroleum spills drift to the polar regions, freeze into the ice and change and darken the albedo, or reflectivity, of the ice, which hastens the eventual thawing of the glaciers. Human forces, it is clear, are the primary shaper now of Earth's climate and geological makeup.

The Cyber Revolution has many aspects, but it is essentially a time of technological revolution. The world as we know it in terms of its politics, economy, social and cultural intercourse, and especially its technology, are changing, and as a consequence Earth's environment increasingly seems at risk. In short, humans are not only polluting the world and consuming its resources but potentially creating such massive changes to its environmental health that the biosphere we live in seems to be endangered.

There are all sorts of ideas about what can be done to rescue the situation – cleaner energy, recycling of resources, controlling the growth of human population and more. There are even more extreme ideas such as to paint the world's clouds white to reflect the Sun's radiation, or even to create a solar shield at Lagrange Point One. This shield might be designed not only to protect Earth from solar storms that could wipe out our electrical grids and pipelines, but also to modulate the Sun's rays to lessen solar warming as well.

If there is an existential question of our age, it is whether globally we will rely on technological innovation or reform its collective behavior to cope effectively with the hyper objects that confront humanity and its ultimate survival. These great challenges, which are as closely linked as Siamese twins, are climate change and overpopulation. The odds are that we will have to deal with them both.

The Cyber Revolution, Family Planning and Overpopulation

The focus on family planning has for decades been centered on creating better prophylactics, birth control pills and other technologies that prevent conception and, in some instances, also protect against sexually transmitted disease. The key missing elements are economic, social or cultural incentives, or disincentives, that provide improved motivation to engage in family planning and seek fewer children.

The number one economic and employment issue closely tied to family planning in many developing countries is the perceived need to create a cadre of farmhands. It is quite reasonable to assume that if one could provide aid or loans to support improved automation in developing countries for agriculture and perhaps also for mining this could aid rural family planning and increase productivity as well. Further, new employment options and alternative jobs to farming and mining again create new dynamics that also support smaller families. The bottom line is that there is not only a need for better birth control capabilities but also benefits to family planning and economic rewards for population control. This can mean tax incentives, economic benefits to prospective parents, new job opportunities, or even formal rewards to villages that find ways to limit population growth. The key step forward rests with the recognition that there needs to be more done than just creating better birth control technology. There must be better economic, social and cultural incentives to induce better family planning. It is imperative, for instance, that the Catholic Church considers the economic, environmental, health and ethical implications of its admonition to its believers to not practice birth control.

The Cyber Revolution, Super-Automation, and Sustainability Within the Fourth Wave Economy

If there are basic and even vital elements of the Cyber Revolution these changes involve work and the striving toward objectives. It was tool-making that differentiated *Homo sapiens* as a species; the idea that humans have always thought of how to create a better existence for themselves and their tribes that has governed the evolution of human civilization. We are unique as a species in our setting of longer-term goals, in the creation of tools and in the setting up of organized units to strive for those objectives. It has been striving and tool-making that have made our species unique.

As far as we know we are the only species that has strategic plans, religions, state governments and vast industrial organizations that can manufacture complex instruments. Central to thousands of years of evolutionary progress has been work. For many, many years it was hunting and gathering. Then, about 10,000 years ago it was farming and the creation of towns and villages. Then came the renaissance and the rise of learning, education, industry and manufacturing. Today, in our post-industrial society, services are becoming predominant, but still working for a living and the setting of goals in life are central to what people do all around the world.

But what if the rise of super-automation, AI-enabled software, smart bots, proto-brains and a breakthrough called the Singularity changes all that? What if work is no longer central to survival? What if machines can farm our lands and reap our crops? What if machines can produce all our cars, our houses and all the thing-a-ma-jigs and doodads that we need? What if smart bots can drive our cars, buses, trucks, trains and aircraft and do so more safely than humans can? What if super-automation can carry out all the tasks that represent work and labor for 99% of all of humanity as work is defined today? What is it that people are supposed to do? Who decides who gets paid a living wage in such a society? If it is the machines that create the wealth and sustain humanity, why does anyone need to work? Does society become a vast army of unemployed proletariat consumers with no purpose in life similar to the hellish life described by Kurt Vonnegut Jr. in *Player Piano*? Or do we morph into a blessed society of near nirvana. Do the faded 1950s dreams of John Kenneth Galbraith, as set forth in *The Affluent Society,* actually become reality? Do the recent aspirations of Peter Diamandis in *Abundance: The Future Is Better*

Than You Think It Is allow us to realize our potential? Are we indeed able to follow the dreams of Elon Musk to cope with climate change, to go and settle Mars and become visionary space travelers, and much more?

This is the great puzzle of our time. There are ideas out there as to how we can adapt to this rapid change that the Cyber Revolution is thrusting into our lives. Are we about to face a fundamental shift in the definition of work and what the goal of human existence should be? Can capitalism, which is based on ideas of work, capital spending, and markets adjust to a world where the old equations no longer work?

Professors Erik Brynjolfsson and Andrew McAfee of the Sloan School of Management at M.I.T. have carried out research on capital markets, worker productivity and employment ever since the end of World War II. For decades everything worked in harmony. New technology, automation, and productivity gains proceeded apace. These factors combined to increase industrial output and profits, and wealth increased and led to the creation of more jobs and more production in an unending circle. But the correlation between productivity and employment that marched in harmony together from the 1940s through the end of the twentieth century are now decoupling. Automation that has become super-automation is now replacing jobs in agriculture, manufacturing and services. Super-automation is no longer leading to the creation of new jobs. Indeed the new jobs that are being produced are low-paying positions such as workers at fast-food restaurants, pre-school aids, etc. The use of technology and super-automation is no longer used to generate wealth that can create jobs but simply to replace the workers.

Rapid technological change has thus been destroying jobs faster than it is creating them, contributing to the stagnation of median income and the growth of inequality in the United States. And, it is likely that something similar is happening in other technologically advanced countries.[5]

According the U. S. Bureau of Labor Statistics, workers' share of U. S. industrial output in terms of income was at a low of 49.5% in 1929, but then rose to a high of 59% in 1970. Since 1970, however, it has been steadily declining to 52% today.[6]

The real question is whether we will find new answers in time. Will entirely new enterprises, such as creating a magnetic barrier out in space to protect Earth from solar storms help? Will we create a similar magnetic barrier to stop the Sun from stripping away the atmosphere of Mars? Such terraforming activities might allow the Red Planet to someday become both blue with water and green with vegetation. We may easily develop the technological knowhow if we have the will.

Will we deploy new green energy systems on Earth, and out in space, to cope with climate change? Will we able to address global needs in transportation, housing and energy through the use of innovative 3D or additive manufacturing systems? Will many new disruptive industries based on the efficiencies of the Internet create enough jobs to offset the overall employment losses due to super-automation? Will

---

[5] David Rotman, "How Technology Is Destroying Jobs" MIT Technology Review, June 12, 2013 https://www.technologyreview.com/s/515926/how-technology-is-destroying-jobs/

[6] Michael Jones, "Yes, the robots will steal our jobs" www.washingtonpost.com/posteverything/ wp/2016/2. Feb, 2016.

new taxes on automation and proto-brain technology be sufficient to provide for the cost of retraining workers for new types of jobs in coming years? At the most profound level, will super-automation force us to restructure the concept of work and compensation during the course of the twenty-first century? Diagnosing the problem associated with super-automation is much easier than finding the answers.

What is almost startlingly true is that the answers related to the use of technology and super-automation will be hugely different in places, as between the Global North and Global South, i.e., the developed and developing countries. There remains a large income, education, healthcare and digital divide gap that exists between the wealthiest and poorest countries. According to the U. N. Facts and Figures produced for the Sustainable Development Goals there are still 800 million people who survive on less than $1.25 (U. S.) a day. Thus great poverty concerns remain in Africa, Asia and Australasia, the Caribbean and South and Central America.

The issues of work, living wage guarantees and super-automation that are beginning to arise in developed economies remain largely foreign to developing countries, but ultimately these issues will still affect the global economy and compensation systems and jobs even in much poorer countries. The great challenge is to make the benefits of smart machines something that can provide aid and relief to the entire world. Such improvements, including more productive use of smart machines, cannot be accomplished with any meaningful results for the developing world without significant change. This change would entail truly significant new capital investments (as in trillions of dollars) and a reduction in population. Such an effort could be, indeed should be, a part of a global effort to confront climate change, bring clean energy to Third World countries, and take seriously the U. N. Sustainable Development Goals.

The Cyber Revolution and the Smart City

The future of the world hinges in large part on the future of our cities. Just a few facts from the U. N. web page regarding the goal to create sustainable cities underscores how critical successful reform of urban planning will be to our achieving success on a number of fronts. Here are just four critical facts[7]:

- By 2030, 60% or more of the world's population will live in urban areas.
- 95% of urban expansion in the next decades will take place in the developing world.
- 828 million people live in slums today, and the number keeps rising.
- The world's cities occupy just 3% of Earth's land but account for 60–80% of energy consumption and 75% of carbon emissions.

---

[7] United Nations Sustainable Development Goals, Targets and Facts and Figures, https://www.un.org/sustainabledevelopment/development-agenda/

These 'facts,' which are based on well-researched projections or detailed demographic studies, suggest that cities, and especially cities in developing countries, are where many key reforms must be made. It means that it is in cities where population growth must be stemmed. It is there that jobs must be found. It is there that key global climate change issues must be addressed. It is there where carbon-based greenhouse gas emissions must be cut. Currently we are falling way short of what must be done. The targets that were set for cities for 2020 are not even close to being met.

Just one of the goals for sustainable cities for 2020 illustrates the point. This goal sets the following targets, none of which will be achieved: "By 2020, substantially increase the number of cities and human settlements adopting and implementing integrated policies and plans towards inclusion, resource efficiency, mitigation and adaptation to climate change, resilience to disasters, and develop and implement, in line with the Sendai Framework for Disaster Risk Reduction 2015–2030, holistic disaster risk management at all levels."[8]

It would be lucky if one city in a hundred would have adopted policies and plans that could seriously address all of the elements in this goal.

There are many things that city planners and officials could do to address what is a mounting urban crisis. They could set goals to create clean energy plans that set specific targets for 20–40 years in the future. They could undertake efforts to create meta-cities on the perimeter of giant megacities in order to decrease intense overcrowding. There could be a major involvement with the citizenry to create a vision of what they want to have their city become in 20–40 years. Once that vision is in place they could undertake to create the transportation, utilities, broadband communications, development and zoning master plans, the education and healthcare systems and protective systems needed to get from point A (i.e.. the here and now) to point B (i.e. the vision for the future).

There is now a guidebook for how to undertake planning and goal-setting to achieve a smart city. This book, by Pelton and Singh, sets forth the key steps to undertake and the traps to avoid. Unfortunately cyber-security and protective strategies against disaster are increasingly important in the age of cybersecurity.[9]

The Cyber Revolution and Education, Training and Healthcare

Many people see the advent of tele-education, tele-training and tele-health as all about cost-cutting and economic efficiency. Clearly these systems can eliminate the cost of bricks and mortar, and share medical, health and educational information more widely and efficiently, thereby reducing costs. But the gains can go far beyond cost reductions. The results are impressive.

---

[8] Ibid.

[9] Joseph Pelton and Indu Singh, *Smart Cities of Today and Tomorrow* (2018) Springer Press, N. Y.

Across the United States [and the rest of the world too], *on-line education is booming. Sixth-through 12th graders enrolled in Florida's largest full-time virtual high school completed more than 44,000 semesters of classwork last year. In Kansas, virtual school enrollment grew 100-fold between 1999 and 2014, from about 60 students to more than 6000.......Thankfully we've begun to appreciate that students aren't stamped from a single mold. Some do their best learning at their own pace and rhythm.*[10]

The gains in online health and medical care have been even more impressive. The ability of AI-enabled software and big data analytics to diagnose illnesses can have great impact in rural and remote areas and allow nurse practitioners to provide quality medical care, so as to reduce costs and deliver care to a much wider circle of people. Exactly how all of this will play out in the future is far from clear.

Some tend to envision a future where medical care is highly automated. Patients, in this future world, would describe their systems to a proto-brain doctor. This would be followed by blood, urine, or other tests that would be instantly processed and then drugs distributed via vending machines with charges going to credit card accounts. The cost of this service would be minimal, and this might be essentially free in a world where socialized medicine and education would be a right extended to all citizens. Others tend to see healthcare and education as something that remains in more limited supply in capitalist nations. In countries such as the United States, the cost of drugs and medical care would remain much higher due to regulatory restrictions, limitations in the number of medical and healthcare practitioners authorized to provide medical services, and patents enforced by pharmaceutical companies.

The real question is whether such starkly different systems can co-exist over the longer term. If one country bordering another provides educational and health services at virtually no cost and of high quality while the other country maintains extremely high costs and limited access, it is hard to imagine that migration from one country to the other would not ensue. This is to say that the Cyber Revolution is almost viral in its impact and ability to spread.

The real question ultimately will be about the relationship of the coming Cyber Revolution and its impact on the nation state and ability to spread across the various countries of the world. The Internet may well be regarded as the first phase of the Cyber Revolution, with companies such as Amazon and Ali Babi offering to a world market the ability to purchase goods – and even services – at lower and lower cost. The rise of disruptive companies such as Uber and Lyft, Airbnb, Fidelity, and Schwab, might be seen as the second tier of this global revolution. If and when the technology and service offering rises to the level that one can obtain low-cost and quality services such as education and healthcare via the Internet on a global scale, the whole concept of nation-state and national borders may come into question. The problem is that one must never confuse quality social services with more cost effective education and healthcare services. They are simply not the same.

---

[10] David von Drehle, "Going to School No Longer Means Going to School" *Washington Post*, May 16, 2018 p. A 15.

Economic and Political Reform

In the age of Google, Yahoo, Amazon, Uber, Ali Baba, Facebook, streaming media and social media there comes a time when the force of Internet connectivity becomes incredibly powerful. At Intelsat in the 1980s, we created a new tariff to sell satellite transponders to developing countries for telephone, data and television services. The first such satellite service in Africa was for Algeria. For centuries the markets in regional capitals had always closed at sundown. Yet in a few weeks of satellite television broadcasts being distributed to these towns, the markets began to close at 5:00 p. m. because that was when the television broadcasts began. Today, in the age of the Internet, borders for trade, the exchange of information, news, sports, and even cultural practices and pornography have broken down more and more.

A really profound question will be posed by the new low-Earth orbit satellite constellations such as those now planned by One Web, Telesat Canada, Boeing, Space X, and China. How will these systems change our world? How will low-cost broadband communications and especially Internet access change social, economic and political behavior in underserved areas of Asia, Africa, and South and Central America? It is no accident that Gregg Wyler, who is the 'father' of One Web, first started a satellite system called O3b that stood for the 'Other Three Billion,' or the people in the equatorial regions of the world that are largely the ones still not connected to the web. Wyler's vision was to bring connectivity and healthcare and education to this underserved part of the world. Today Google, Apple, and Amazon see the potential to expand the markets they serve. The nation states of the world have not recognized that once the world is connected to the web, their world may be changed forever.

How does sending targeted bot messaging via social media impact democratic elections? What impact does seeing streamed news, sports and entertainment from overseas do to authoritarian states and their control of their citizenry? How do people react when they can see a different world, where freedom and liberty of association and expression are available to people overseas? Does the worldwide spread of the Internet and the Internet of Things mean the end of personal privacy? Do new consumer privacy laws truly allow personal privacy to exist in the age of social media? Do regulatory privacy guarantees (see Appendices) really work or are they merely window dressing?

The first adjustments that nation-states will seek to make relate to communications and Internet access. But many feel that this is too little, too late. Two of the biggest violators of international standards against jamming of satellites today, according to Eutelsat, the large European satellite operator, are Iran and Syria. These countries have tried, largely unsuccessfully, to prevent video from reaching their citizens and to learn what is happening in the rest of the world. In the world of Internet and worldwide media it is increasingly difficult to prevent people knowing how the rest of the world lives.

However, the electronic access that comes via the Internet, radio and television broadcasts, and satellite connections is only the first element of change. The rapid evolution of AI-based systems will penetrate everywhere in an increasingly global marketplace. Disruptions of global markets will begin to occur all over. Some may see these changes as good, others bad. The questions are everywhere. Can print newspapers selling subscriptions to subscribers continue to compete with free online news provided via social media? Can conventional doctors and hospitals continue to offer tremendously expensive services? Can pharmaceutical companies continue to sell extremely high cost drugs, if there are online services and proto-brain-based AI systems that can offer alternative capabilities? Can nations control their borders and impose international duties on systems and services that can be offered online? In an age of 3D printing and additive manufacturing, will nations be able to control illegal printing of products such as guns and other contraband items? Unfortunately the latest attempts to use U. S. courts to prevent the 3D printing of plastic guns and the smart milling of automatic assault weapons in July 2018 suggests the answer is no. Stopping the spread of information on the web about weapons and how to build them seems much like trying to stick a finger in a dike that is riddled with dozens of holes.[11]

The biggest challenge of all will be the ultimate advent of the Fourth Wave economy that will increasingly redefine conventional concepts that governments and economists have now depended on for centuries. In the age of the Singularity, what will be the new meaning of work, or compensation or salary for work performed? Basic concepts of wealth, living wages, retirement, taxation, government spending and economic incentives may have to be reconceived. Will art and culture become more important? How do people in a world that essentially runs on super-automation interact with a world that is generally reliant on human labor? These and hundreds of other questions now swirl around the world of smart bots and AI proto-brains. We don't have to worry about these things yet. All in good time we will sort these things out with meaningful answers. But the nagging question comes back: "Will we?"

Next Steps

If there is one document that reveals the many dilemmas that the modern world faces, it is the U. N. web pages that set forth the Sustainable Development Goals, targets that have been set, and facts and figures on which this complicated document is based. The one thing that is abundantly clear is that the world is not ready for the Cyber Revolution. The only good news is that the world was not ready for the world of the Internet but somehow we are muddling through and surviving it. Young people see it as a new ecosphere to survive and even thrive using. Millennials and younger have largely embraced it.

The world of radio and television tended to be a place that held communities and nations together. These broadcast media for decades created what was largely a cohesive social glue. The world of Internet and, in a way, its competitor, the 24-h

---

[11] Dan Freedman "Court blocks 3D gun printing," *CT Post*, July 31, 2018. https://www.ctpost.com/politics/article/Fight-to-block-3D-gun-printing-goes-down-to-wire-13121383.php

cable news channels and special interest cable programming such as music video channels, religious channels and now even a streamed National Rifle Association video channel, have increasingly defined new media-based groups that are divided into political or cultural interest groups. Internet chat groups, social media, right- and left-wing news channels and more help to create what are essentially warring political and sectarian groups. This new world has become dangerously splintered, and most Internet and cable-based programming is driving splinter groups further apart.

The world of Internet, broadband mobile communications, AI-based software, and Internet commerce and social media is, in a word, complicated. The Cyber Revolution will provide us powerful new tools to help attack the problems of tomorrow.

Conclusion

In order to reach the great potential of a truly abundant economy, there is only one viable pathway forward. We must create a greener and more sustainable world. This will require the focused use of technology to prioritize sustainability and invest in creating new industries that can actually help us address the really big challenges that will involve creating viable new green industries that let us create new jobs and profits but also be saving Earth as a viable biosphere for life as we know it. This means a host of new sustainable industries that include electric cars, better solar energy systems with longer lives, ocean thermal energy conversion plants, desalinization plants, solar power satellites, new and better birth control products and better family planning, and thousands of new products and services that replace the outmoded and dirty carbon-based industrial economy of the past. This means not only developing new technology and new green industries but the creation of political, economic and regulatory systems that hastens the transition from the old to the new.

It is important to help society recognize the danger represented by the so-called hyper-objects that actually threaten the survival of our modern society and our vital infrastructure. It means giving priority to addressing the very real problems of overpopulation, coping with the unprecedented growth of megacities, attacking climate change and channeling new technology toward the most important of the survivability goals. These are now absolute priority goals.

We must look to creating a Fourth Wave economy that is circular rather than disposable in nature. We need to re-tune the forces of democratic capitalism to recognize that simply increasing throughput and making a profit will be a dead end for the human race. The worlds "dead" and "end" will become literally true for our species if we do not bend the will of the Cyber Revolution to human survivability goals. The next few decades of the twenty-first century represent a potentially very narrow passageway to the future. The difference could be a glorious and abundant future, or the end of the line for humanity and our fellow passengers on our 6-sextillion-ton spacecraft.

If we take the wrong road forward we could very well become like lobsters in a pot that do not recognize that the water we inhabit is gradually heating up, and in time we will eventually and very surely be cooked to death. Change and reform are absolutely key to the future. We have to channel the great potential of the Cyber Revolution in the direction of survival and sustainability. There is very little room for error.

Correction to: Preparing for the Next Cyber Revolution

Correction to:
J. N. Pelton, *Preparing for the Next Cyber Revolution*,
https://doi.org/10.1007/978-3-030-02137-5

This book was inadvertently published without updating the corrections to the chapters noted below.

In Chapter 1, page 12, the text in the section "A Quick Tour Guide of the Cyber Revolution" has been replaced.

In Chapter 1, page 8, the sentence starting with "Ray Kurzweil won .... in 2015" has been corrected to read as "Greg Wyler won the Arthur C. Clarke Innovators award in 2015."

In Chapter 4, page 67, the quote at the commencement of the chapter has been corrected to read as: "A city is defined by its sense of community." Paul and Percival Goodman

---

The updated online versions of these chapters can be found at
https://doi.org/10.1007/978-3-030-02137-5_1
https://doi.org/10.1007/978-3-030-02137-5_4
https://doi.org/10.1007/978-3-030-02137-5

© Springer Nature Switzerland AG 2019   C1
J. N. Pelton, *Preparing for the Next Cyber Revolution*,
https://doi.org/10.1007/978-3-030-02137-5_11

Appendix 1: The U. N. Sustainable Development Goals

As a follow up to the United Nations' adoption of its Millennium Development Goals, the General Assembly on January 2, 2016, adopted s new set of Sustainable Development Goals as provided below. This appendix, in addition to providing the official brief explanation of the 17 Goals, also provides the U. N.-approved targets for each goal as well as related facts and figures associated with these objectives that supports the rationale for their adoption.

Goal 1: End Poverty

Eradicating poverty in all its forms remains one of the greatest challenges facing humanity. While the number of people living in extreme poverty dropped by more than half between 1990 and 2015 – from 1.9 billion to 836 million – too many are still struggling for the most basic human needs.

Globally, more than 800 million people are still living on less than US$1.25 a day, many lacking access to adequate food, clean drinking water and sanitation. Rapid economic growth in countries like China and India has lifted millions out of poverty, but progress has been uneven. Women are more likely to live in poverty than men due to unequal access to paid work, education and property.

Progress has also been limited in other regions, such as South Asia and sub-Saharan Africa, which account for 80% of those living in extreme poverty. New threats brought on by climate change, conflict and food insecurity, mean even more work is needed to bring people out of poverty.

The SDGs are a bold commitment to finish what we started, and end poverty in all forms and dimensions by 2030. This involves targeting the most vulnerable, increasing access to basic resources and services, and supporting communities affected by conflict and climate-related disasters.

© Springer Nature Switzerland AG 2019   171
J. N. Pelton, *Preparing for the Next Cyber Revolution*,
https://doi.org/10.1007/978-3-030-02137-5

Goal 1 Targets
- By 2030, reduce at least by half the proportion of men, women and children of all ages living in poverty in all its dimensions according to national definitions.
- Implement nationally appropriate social protection systems and measures for all, including floors, and by 2030 achieve substantial coverage of the poor and the vulnerable.
- By 2030, ensure that all men and women, in particular the poor and the vulnerable, have equal rights to economic resources, as well as access to basic services, ownership and control over land and other forms of property, inheritance, natural resources, appropriate new technology and financial services, including microfinance.
- By 2030, build the resilience of the poor and those in vulnerable situations and reduce their exposure and vulnerability to climate-related extreme events and other economic, social and environmental shocks and disasters.
- Ensure significant mobilization of resources from a variety of sources, including through enhanced development cooperation, in order to provide adequate and predictable means for developing countries, in particular least developed countries, to implement. Programmes and policies to end poverty in all its dimensions.
- Create sound policy frameworks at the national, regional and international levels, based on pro-poor and gender-sensitive development strategies, to support accelerated investment in poverty eradication actions.

Facts and Figures
- 836 million people still live in extreme poverty.
- About one in five persons in developing regions lives on less than US$1.25 per day.
- The overwhelming majority of people living on less than $1.25 a day belong to two regions: Southern Asia and sub-Saharan Africa.
- High poverty rates are often found in small, fragile and conflict-affected countries.
- One in four children under age five in the world has inadequate height for his or her age.
- Every day in 2014, 42,000 people had to abandon their homes to seek protection due to conflict.

Goal 2: Zero Hunger

Rapid economic growth and increased agricultural productivity over the past two decades have seen the number of undernourished people drop by almost half. Many developing countries that used to suffer from famine and hunger can now meet the nutritional needs of the most vulnerable. Central and East Asia, Latin America and the Caribbean have all made huge progress in eradicating extreme hunger.

These are all huge achievements in line with the targets set out by the first Millennium Development Goals. Unfortunately, extreme hunger and malnutrition remain a huge barrier to development in many countries. Seven hundred and ninety five million people are estimated to be chronically undernourished as of 2014, often as a direct consequence of environmental degradation, drought and loss of biodiversity. Over 90 million children under the age of five are dangerously underweight. And one person in every four still goes hungry in Africa.

The SDGs aim to end all forms of hunger and malnutrition by 2030, making sure all people – especially children – have access to sufficient and nutritious food all year round. This involves promoting sustainable agricultural practices: supporting small scale farmers and allowing equal access to land, technology and markets. It also requires international cooperation to ensure investment in infrastructure and technology to improve agricultural productivity. Together with the other goals set out here, we can end hunger by 2030.

There is an imperative today to foster sustainable development. A vision for what this encapsulates is laid out in the new sustainable development agenda that aims to end poverty, promote prosperity and people's well-being while protecting the environment by 2030. As the UN's Development arm, UNDP has a key role to play in supporting countries to make this vision a reality – putting societies on a sustainable development pathway, managing risk and enhancing resilience, and advancing prosperity and well-being.

Goal 2 Targets
- By 2030, end hunger and ensure access by all people, in particular the poor and people in vulnerable situations, including infants, to safe, nutritious and sufficient food all year round
- By 2030, end all forms of malnutrition, including achieving, by 2025, the internationally agreed targets on stunting and wasting in children under 5 years of age, and address the nutritional needs of adolescent girls, pregnant and lactating women and older persons.
- By 2030, double the agricultural productivity and incomes of small-scale food producers, in particular women, indigenous peoples, family farmers, pastoralists and fishers, including through secure and equal access to land, other productive resources and inputs, knowledge, financial services, markets and opportunities for value addition and non-farm employment.
- By 2030, ensure sustainable food production systems and implement resilient agricultural practices that increase productivity and production, that help maintain ecosystems, that strengthen capacity for adaptation to climate change, extreme weather, drought, flooding and other disasters and that progressively improve land and soil quality.
- By 2020, maintain the genetic diversity of seeds, cultivated plants and farmed and domesticated animals and their related wild species, including through soundly managed and diversified seed and plant banks at the national, regional and international levels, and promote access to and fair and equitable sharing of benefits arising from the utilization of genetic resources and associated traditional knowledge, as internationally agreed.

- Increase investment, including through enhanced international cooperation, in rural infrastructure, agricultural research and extension services, technology development and plant and livestock gene banks in order to enhance agricultural productive capacity in developing countries, in particular least developed countries.
- Correct and prevent trade restrictions and distortions in world agricultural markets, including through the parallel elimination of all forms of agricultural export subsidies and all export measures with equivalent effect, in accordance with the mandate of the Doha Development Round.
- Adopt measures to ensure the proper functioning of food commodity markets and their derivatives and facilitate timely access to market information, including on food reserves, in order to help limit extreme food price volatility.

Facts and Figures
- Globally, one in nine people in the world today (795 million) are undernourished.
- The vast majority of the world's hungry people live in developing countries, where 12.9% of the population is undernourished.
- Asia is the continent with the most hungry people – two thirds of the total. The percentage in southern Asia has fallen in recent years but in western Asia it has increased slightly.
- Southern Asia faces the greatest hunger burden, with about 281 million undernourished people. In sub-Saharan Africa, projections for the 2014–2016 period indicate a rate of undernourishment of almost 23%.
- Poor nutrition causes nearly half (45%) of deaths in children under five – 3.1 million children each year.
- One in four of the world's children suffer stunted growth. In developing countries the proportion can rise to one in three.
- 66 million primary school-age children attend classes hungry across the developing world, with 23 million in Africa alone.
- Agriculture is the single largest employer in the world, providing livelihoods for 40% of today's global population. It is the largest source of income and jobs for poor rural households.

Goal 3: Good Health and Well-Being

Ensure Healthy Live and Promote Well Being

Ensuring healthy lives and promoting the well-being for all at all ages is essential to sustainable development. Significant strides have been made in increasing life expectancy and reducing some of the common killers associated with child and maternal mortality. Major progress has been made on increasing access to clean water and sanitation, reducing malaria, tuberculosis, polio and the spread of HIV/ AIDS. However, many more efforts are needed to fully eradicate a wide range of diseases and address many different persistent and emerging health issues.

Goal 3 Targets
- By 2030, reduce the global maternal mortality ratio to less than 70 per 100,000 live births.
- By 2030, end preventable deaths of newborns and children under 5 years of age, with all countries aiming to reduce neonatal mortality to at least as low as 12 per 1000 live births and under-5 mortality to at least as low as 25 per 1000 live births.
- By 2030, end the epidemics of AIDS, tuberculosis, malaria and neglected tropical diseases and combat hepatitis, water-borne diseases and other communicable diseases.
- By 2030, reduce by one third premature mortality from non-communicable diseases through prevention and treatment and promote mental health and well-being.
- Strengthen the prevention and treatment of substance abuse, including narcotic drug abuse and harmful use of alcohol.
- By 2020, halve the number of global deaths and injuries from road traffic accidents.
- By 2030, ensure universal access to sexual and reproductive health-care services, including for family planning, information and education, and the integration of reproductive health into national strategies and programmes.
- Achieve universal health coverage, including financial risk protection, access to quality essential health-care services and access to safe, effective, quality and affordable essential medicines and vaccines for all.
- By 2030, substantially reduce the number of deaths and illnesses from hazardous chemicals and air, water and soil pollution and contamination.
- Strengthen the implementation of the World Health Organization Framework Convention on Tobacco Control in all countries, as appropriate.
- Support the research and development of vaccines and medicines for the communicable and non-communicable diseases that primarily affect developing countries, provide access to affordable essential medicines and vaccines, in accordance with the Doha Declaration on the TRIPS Agreement and Public Health, which affirms the right of developing countries to use to the full the provisions in the Agreement on Trade Related Aspects of Intellectual Property Rights regarding flexibilities to protect public health, and, in particular, provide access to medicines for all.
- Substantially increase health financing and the recruitment, development, training and retention of the health workforce in developing countries, especially in least developed countries and small island developing States.
- Strengthen the capacity of all countries, in particular developing countries, for early warning, risk reduction and management of national and global health risks.

Facts and Figures

- 17,000 fewer children die each day than in 1990, but more than six million children still die before their fifth birthday each year.
- Since 2000, measles vaccines have averted nearly 15.6 million deaths.
- Despite determined global progress, an increasing proportion of child deaths are in sub-Saharan Africa and Southern Asia. Four out of every five deaths of children under age five occur in these regions.
- Children born into poverty are almost twice as likely to die before the age of five as those from wealthier families.
- Children of educated mothers – even mothers with only primary schooling – are more likely to survive than children of mothers with no education.
- Maternal mortality has fallen by almost 50% since 1990.
- In Eastern Asia, Northern Africa and Southern Asia, maternal mortality has declined by around two-thirds.
- But maternal mortality ratio – the proportion of mothers that do not survive childbirth compared to those who do – in developing regions is still 14 times higher than in the developed region.
- More women are receiving antenatal care. In developing regions, antenatal care increased from 65% in 1990 to 83% in 2012.
- Only half of women in developing regions receive the recommended amount of health care they need.
- Fewer teens are having children in most developing regions, but progress has slowed. The large increase in contraceptive use in the 1990s was not matched in the 2000s
- The need for family planning is slowly being met for more women, but demand is increasing at a rapid pace.
- At the end of 2014, there were 13.6 million people accessing antiretroviral therapy.
- New HIV infections in 2013 were estimated at 2.1 million, which was 38% lower than in 2001.
- At the end of 2013, there were an estimated 35 million people living with HIV.
- At the end of 2013, 240,000 children were newly infected with HIV.
- New HIV infections among children have declined by 58% since 2001.
- Globally, adolescent girls and young women face gender-based inequalities, exclusion, discrimination and violence, which put them at increased risk of acquiring HIV.
- HIV is the leading cause of death for women of reproductive age worldwide.
- E.TB-related deaths in people living with HIV have fallen by 36% since 2004.
- There were 250,000 new HIV infections among adolescents in 2013, two thirds of which were among adolescent girls.
- AIDS is now the leading cause of death among adolescents (aged 10–19) in Africa and the second most common cause of death among adolescents globally.
- In many settings, adolescent girls' right to privacy and bodily autonomy is not respected, as many report that their first sexual experience was forced.

- As of 2013, 2.1 million adolescents were living with HIV.
- Over 6.2 million malaria deaths have been averted between 2000 and 2015, primarily of children under 5 years of age in sub-Saharan Africa. The global malaria incidence rate has fallen by an estimated 37% and the mortality rates by 58%.
- Between 2000 and 2013, tuberculosis prevention, diagnosis and treatment interventions saved an estimated 37 million lives. The tuberculosis mortality rate fell by 45% and the prevalence rate by 41% between 1990 and 2013.

Goal 4: Quality Education

Ensure Inclusive and Quality Education for All and Promote Lifelong Learning

Obtaining a quality education is the foundation to improving people's lives and sustainable development. Major progress has been made towards increasing access to education at all levels and increasing enrolment rates in schools particularly for women and girls. Basic literacy skills have improved tremendously, yet bolder efforts are needed to make even greater strides for achieving universal education goals. For example, the world has achieved equality in primary education between girls and boys, but few countries have achieved that target at all levels of education.

Goal 4 Targets
- By 2030, ensure that all girls and boys complete free, equitable and quality primary and secondary education leading to relevant and Goal-4 effective learning outcomes.
- By 2030, ensure that all girls and boys have access to quality early childhood development, care and preprimary education so that they are ready for primary education.
- By 2030, ensure equal access for all women and men to affordable and quality technical, vocational and tertiary education, including university.
- By 2030, substantially increase the number of youth and adults who have relevant skills, including technical and vocational skills, for employment, decent jobs and entrepreneurship.
- By 2030, eliminate gender disparities in education and ensure equal access to all levels of education and vocational training for the vulnerable, including persons with disabilities, indigenous peoples and children in vulnerable situations.
- By 2030, ensure that all youth and a substantial proportion of adults, both men and women, achieve literacy and numeracy.
- By 2030, ensure that all learners acquire the knowledge and skills needed to promote sustainable development, including, among others, through education for sustainable development and sustainable lifestyles, human rights, gender equality, promotion of a culture of peace and non-violence, global citizenship

and appreciation of cultural diversity and of culture's contribution to sustainable development.
- Build and upgrade education facilities that are child, disability and gender sensitive and provide safe, nonviolent, inclusive and effective learning environments for all.
- By 2020, substantially expand globally the number of scholarships available to developing countries, in particular least developed countries, small island developing States and African countries, for enrolment in higher education, including vocational training and information and communications technology, technical, engineering and scientific programmes, in developed countries and other developing countries.
- By 2030, substantially increase the supply of qualified teachers, including through international cooperation for teacher training in developing countries, especially least developed countries and small island developing states.

Facts and Figures
- Enrolment in primary education in developing countries has reached 91% but 57 million children remain out of school.
- More than half of children that have not enrolled in school live in sub-Saharan Africa.
- An estimated 50% of out-of-school children of primary school age live in conflict-affected areas.
- 103 million youth worldwide lack basic literacy skills, and more than 60% of them are women.

Goal 5: Gender Equality

Achieve Gender Equality and Empower All Women and Girls

While the world has achieved progress towards gender equality and women's empowerment under the Millennium Development Goals (including equal access to primary education between girls and boys), women and girls continue to suffer discrimination and violence in every part of the world.

Gender equality is not only a fundamental human right but a necessary foundation for a peaceful, prosperous and sustainable world.

Providing women and girls with equal access to education, health care, decent work, and representation in political and economic decision-making processes will fuel sustainable economies and benefit societies and humanity at large.

Goal 5 Targets
- End all forms of discrimination against all women and girls everywhere.
- Eliminate all forms of violence against all women and girls in the public and private spheres, including trafficking and sexual and other types of exploitation.

- Eliminate all harmful practices, such as child, early and forced marriage and female genital mutilation.
- Recognize and value unpaid care and domestic work through the provision of public services, infrastructure and social protection policies and the promotion of shared responsibility within the household and the family as nationally appropriate.
- Ensure women's full and effective participation and equal opportunities for leadership at all levels of decision-making in political, economic and public life.
- Ensure universal access to sexual and reproductive health and reproductive rights as agreed in accordance with the Programme of Action of the International Conference on Population and Development and the Beijing Platform for Action and the outcome documents of their review conferences.
- Undertake reforms to give women equal rights to economic resources, as well as access to ownership and control over land and other forms of property, financial services, inheritance and natural resources, in accordance with national laws.
- Enhance the use of enabling technology, in particular information and communications technology, to promote the empowerment of women.
- Adopt and strengthen sound policies and enforceable legislation for the promotion of gender equality and the empowerment of all women and girls at all levels

Facts and Figures
- About two thirds of countries in the developing regions have achieved gender parity in primary education.
- In Southern Asia, only 74 girls were enrolled in primary school for every 100 boys in 1990. By 2012, the enrolment ratios were the same for girls as for boys.
- In sub-Saharan Africa, Oceania and Western Asia, girls still face barriers to entering both primary and secondary school.
- Women in Northern Africa hold less than one in five paid jobs in the non-agricultural sector. The proportion of women in paid employment outside the agriculture sector has increased from 35% in 1990 to 41% in 2015.
- In 46 countries, women now hold more than 30% of seats in national parliament in at least one chamber.

Goal 6: Clean Water and Sanitation

Ensure Access to Water and Sanitation for All

Clean, accessible water for all is an essential part of the world we want to live in. There is sufficient fresh water on the planet to achieve this. But due to bad economics or poor infrastructure, every year millions of people, most of them children, die from diseases associated with inadequate water supply, sanitation and hygiene.

Water scarcity, poor water quality and inadequate sanitation negatively impact food security, livelihood choices and educational opportunities for poor families across the world. Drought afflicts some of the world's poorest countries, worsening hunger and malnutrition.

By 2050, at least one in four people is likely to live in a country affected by chronic or recurring shortages of fresh water.

Goal 6 Targets

- By 2030, achieve universal and equitable access to safe and affordable drinking water for all.
- By 2030, achieve access to adequate and equitable sanitation and hygiene for all and end open defecation, paying special attention to the needs of women and girls and those in vulnerable situations.
- By 2030, improve water quality by reducing pollution, eliminating dumping and minimizing release of hazardous chemicals and materials, halving the proportion of untreated wastewater and substantially increasing recycling and safe reuse globally.
- By 2030, substantially increase water-use efficiency across all sectors and ensure sustainable withdrawals and supply of freshwater to address water scarcity and substantially reduce the number of people suffering from water scarcity.
- By 2030, implement integrated water resources management at all levels, including through trans-boundary cooperation as appropriate.
- By 2020, protect and restore water-related ecosystems, including mountains, forests, wetlands, rivers, aquifers and lakes.
- By 2030, expand international cooperation and capacity-building support to developing countries in water- and sanitation-related activities and programmes, including water harvesting, desalination, water efficiency, wastewater treatment, recycling and reuse technologies
- Support and strengthen the participation of local communities in improving water and sanitation management

Facts and Figures

- 2.6 billion people have gained access to improved drinking water sources since 1990, but 663 million people are still without.
- At least 1.8 billion people globally use a source of drinking water that is fecally contaminated.
- Between 1990 and 2015, the proportion of the global population using an improved drinking water source has increased from 76% to 91%.
- But water scarcity affects more than 40% of the global population and is projected to rise. Over 1.7 billion people are currently living in river basins where water use exceeds recharge.
- 2.4 billion people lack access to basic sanitation services, such as toilets or latrines.
- More than 80% of wastewater resulting from human activities is discharged into rivers or sea without any pollution removal.

- Each day, nearly 1000 children die due to preventable water and sanitation-related diarrhoeal diseases.
- Hydropower is the most important and widely-used renewable source of energy and as of 2011, represented 16% of total electricity production worldwide.
- Approximately 70% of all water abstracted from rivers, lakes and aquifers is used for irrigation.
- Floods and other water-related disasters account for 70% of all deaths related to natural disaster

Goal 7: Affordable and Clean Energy

Ensure Access to Affordable, Reliable, Sustainable and Modern Energy for All

Energy is central to nearly every major challenge and opportunity the world faces today. Be it for jobs, security, climate change, food production or increasing incomes, access to energy for all is essential.

Sustainable energy is opportunity – it transforms lives, economies and the planet.

UN Secretary-General Ban Ki-moon is leading a Sustainable Energy for All initiative to ensure universal access to modern energy services, improve efficiency and increase use of renewable sources.

Goal 7 Targets
- By 2030, ensure universal access to affordable, reliable and modern energy services.
- By 2030, increase substantially the share of renewable energy in the global energy mix.
- By 2030, double the global rate of improvement in energy efficiency.
- By 2030, enhance international cooperation to facilitate access to clean energy research and technology, including renewable energy, energy efficiency and advanced and cleaner fossil-fuel technology, and promote investment in energy infrastructure and clean energy technology.
- By 2030, expand infrastructure and upgrade technology for supplying modern and sustainable energy services for all in developing countries, in particular least developed countries, small island developing States, and land-locked developing countries, in accordance with their respective programmes of support.

Facts and Figures
- One in five people still lacks access to modern electricity.
- 3 billion people rely on wood, coal, charcoal or animal waste for cooking and heating.
- Energy is the dominant contributor to climate change, accounting for around 60% of total global greenhouse gas emissions.
- Reducing the carbon intensity of energy is a key objective in long-term climate goals.

Goal 8: Decent Work and Economic Growth

Promote Inclusive and Sustainable Economic Growth, Employment and Decent Work for All

Roughly half the world's population still lives on the equivalent of about US$2 a day. And in too many places, having a job doesn't guarantee the ability to escape from poverty. This slow and uneven progress requires us to rethink and retool our economic and social policies aimed at eradicating poverty.

A continued lack of decent work opportunities, insufficient investments and under-consumption lead to an erosion of the basic social contract underlying democratic societies: that all must share in progress. The creation of quality jobs will remain a major challenge for almost all economies well beyond 2015.

Sustainable economic growth will require societies to create the conditions that allow people to have quality jobs that stimulate the economy while not harming the environment. Job opportunities and decent working conditions are also required for the whole working age population.

Goal 8 Targets

- Sustain per capita economic growth in accordance with national circumstances and, in particular, at least 7% gross domestic product growth per annum in the least developed countries.
- Achieve higher levels of economic productivity through diversification, technological upgrading and innovation, including through a focus on high-value added and labour-intensive sectors.
- Promote development-oriented policies that support productive activities, decent job creation, entrepreneurship, creativity and innovation, and encourage the formalization and growth of micro-, small- and medium-sized enterprises, including through access to financial services.
- Improve progressively, through 2030, global resource efficiency in consumption and production and endeavour to decouple economic growth from environmental degradation, in accordance with the 10-year framework of programmes on sustainable consumption and production, with developed countries taking the lead.
- By 2030, achieve full and productive employment and decent work for all women and men, including for young people and persons with disabilities, and equal pay for work of equal value.
- By 2020, substantially reduce the proportion of youth not in employment, education or training.
- Take immediate and effective measures to eradicate forced labour, end modern slavery and human trafficking and secure the prohibition and elimination of the worst forms of child labour, including recruitment and use of child soldiers, and by 2025 end child labour in all its forms.

- Protect labour rights and promote safe and secure working environments for all workers, including migrant workers, in particular women migrants, and those in precarious employment.
- By 2030, devise and implement policies to promote sustainable tourism that creates jobs and promotes local culture and products.
- Strengthen the capacity of domestic financial institutions to encourage and expand access to banking, insurance and financial services for all.
- Increase Aid for Trade support for developing countries, in particular least developed countries, including through the Enhanced Integrated Framework for Trade-Related Technical Assistance to Least Developed Countries.
- By 2020, develop and operationalize a global strategy for youth employment and implement the Global Jobs Pact of the International Labour Organization.

Facts and Figures
- Global unemployment increased from 170 million in 2007 to nearly 202 million in 2012, of which about 75 million are young women and men.
- Nearly 2.2 billion people live below the US$2 poverty line and that poverty eradication is only possible through stable and well-paid jobs.
- 470 million jobs are needed globally for new entrants to the labour market between 2016 and 2030.

Goal 9: Industry, Innovation and Infrastructure

Build Resilient Infrastructure, Promote Sustainable Industrialization and Foster Innovation

Investments in infrastructure – transport, irrigation, energy and information and communication technology – are crucial to achieving sustainable development and empowering communities in many countries. It has long been recognized that growth in productivity and incomes, and improvements in health and education outcomes require investment in infrastructure.

Inclusive and sustainable industrial development is the primary source of income generation, allows for rapid and sustained increases in living standards for all people, and provides the technological solutions to environmentally sound industrialization.

Technological progress is the foundation of efforts to achieve environmental objectives, such as increased resource and energy-efficiency. Without technology and innovation, industrialization will not happen, and without industrialization, development will not happen.

Goal 9 Targets
- Develop quality, reliable, sustainable and resilient infrastructure, including regional and trans-border infrastructure, to support economic development and human well-being, with a focus on affordable and equitable access for all.
- Promote inclusive and sustainable industrialization and, by 2030, significantly raise industry's share of employment and gross domestic product, in line with national circumstances, and double its share in least developed countries.
- Increase the access of small-scale industrial and other enterprises, in particular in developing countries, to financial services, including affordable credit, and their integration into value chains and markets.
- By 2030, upgrade infrastructure and retrofit industries to make them sustainable, with increased resource-use efficiency and greater adoption of clean and environmentally sound technologies and industrial processes, with all countries taking action in accordance with their respective capabilities.
- Enhance scientific research, upgrade the technological capabilities of industrial sectors in all countries, in particular developing countries, including, by 2030, encouraging innovation and substantially increasing the number of research and development workers per one million people and public and private research and development spending.
- Facilitate sustainable and resilient infrastructure development in developing countries through enhanced financial, technological and technical support to African countries, least developed countries, landlocked developing countries and small island developing States 18.
- Support domestic technology development, research and innovation in developing countries, including by ensuring a conducive policy environment for, inter alia, industrial diversification and value addition to commodities.
- Significantly increase access to information and communications technology and strive to provide universal and affordable access to the Internet in least developed countries by 2020.

Facts and Figures
- Basic infrastructure like roads, information and communication technologies, sanitation, electrical power and water remains scarce in many developing countries.
- About 2.6 billion people in the developing world are facing difficulties in accessing electricity full time.
- 2.5 billion people worldwide lack access to basic sanitation and almost 800 million people lack access to water, many hundreds of millions of them in Sub Saharan Africa and South Asia.
- 1–1.5 billion people do not have access to reliable phone services.
- Quality infrastructure is positively related to the achievement of social, economic and political goals.
- Inadequate infrastructure leads to a lack of access to markets, jobs, information and training, creating a major barrier to doing business.
- Undeveloped infrastructures limits access to health care and education.

- For many African countries, particularly the lower-income countries, the existent constraints regarding infrastructure affect firm productivity by around 40%.
- Manufacturing is an important employer, accounting for around 470 million jobs worldwide in 2009 – or around 16% of the world's workforce of 2.9 billion. In 2013, it is estimated that there were more than half a billion jobs in manufacturing.
- Industrialization's job multiplication effect has a positive impact on society. Every one job in manufacturing creates 2.2 jobs in other sectors.
- Small and medium-sized enterprises that engage in industrial processing and manufacturing are the most critical for the early stages of industrialization and are typically the largest job creators. They make up over 90% of business world-wide and account for between 50% and 60% of employment.
- In countries where data are available, the number of people employed in renewable energy sectors is presently around 2.3 million. Given the present gaps in information, this is no doubt a very conservative figure. Because of strong rising interest in energy alternatives, the possible total employment for renewables by 2030 is 20 million jobs.
- Least developed countries have immense potential for industrialization in food and beverages (agro-industry), and textiles and garments, with good prospects for sustained employment generation and higher productivity.
- Middle-income countries can benefit from entering the basic and fabricated metals industries, which offer a range of products facing rapidly growing international demand.
- In developing countries, barely 30% of agricultural production undergoes industrial processing. In high-income countries, 98% is processed. This suggests that there are great opportunities for developing countries in agribusiness.

Goal 10: Reduced Inequalities

Reduce Inequality Within and Among Countries

The international community has made significant strides towards lifting people out of poverty. The most vulnerable nations – the least developed countries, the land-locked developing countries and the small island developing states – continue to make inroads into poverty reduction. However, inequality still persists and large disparities remain in access to health and education services and other assets.

Additionally, while income inequality between countries may have been reduced, inequality within countries has risen. There is growing consensus that economic growth is not sufficient to reduce poverty if it is not inclusive and if it does not involve the three dimensions of sustainable development – economic, social and environmental.

To reduce inequality, policies should be universal in principle paying attention to the needs of disadvantaged and marginalized populations.

Goal 10 Targets
- By 2030, progressively achieve and sustain income growth of the bottom 40% of the population at a rate higher than the national average.
- By 2030, empower and promote the social, economic and political inclusion of all, irrespective of age, sex, disability, race, ethnicity, origin, religion or economic or other status.
- Ensure equal opportunity and reduce inequalities of outcome, including by eliminating discriminatory laws, policies and practices and promoting appropriate legislation, policies and action in this regard.
- Adopt policies, especially fiscal, wage and social protection policies, and progressively achieve greater equality.
- Improve the regulation and monitoring of global financial markets and institutions and strengthen the implementation of such regulations.
- Ensure enhanced representation and voice for developing countries in decision-making in global international economic and financial institutions in order to deliver more effective, credible, accountable and legitimate institutions.
- Facilitate orderly, safe, regular and responsible migration and mobility of people, including through the implementation of planned and well-managed migration policies.
- Implement the principle of special and differential treatment for developing countries, in particular least developed countries, in accordance with World Trade Organization agreements.
- Encourage official development assistance and financial flows, including foreign direct investment, to States where the need is greatest, in particular least developed countries, African countries, small island developing States and landlocked developing countries, in accordance with their national plans and programmes.
- By 2030, reduce to less than 3% the transaction costs of migrant remittances and eliminate remittance corridors with costs higher than 5%.
- More must be done to stop babies from dying the day they are born, United Nations agencies said in a new report issued Thursday, which argued that life-saving know-how and technologies must be made readily available – particularly in Southern Asia and sub-Saharan Africa – where they are most needed.

Facts and Figures
- On average – and taking into account population size – income inequality increased by 11% in developing countries between 1990 and 2010.
- A significant majority of households in developing countries – more than 75% of the population – are living today in societies where income is more unequally distributed than it was in the 1990s.
- Evidence shows that, beyond a certain threshold, inequality harms growth and poverty reduction, the quality of relations in the public and political spheres and individuals' sense of fulfilment and self-worth.
- There is nothing inevitable about growing income inequality; several countries have managed to contain or reduce income inequality while achieving strong growth performance.

- Income inequality cannot be effectively tackled unless the underlying inequality of opportunities is addressed.
- In a global survey conducted by UN Development Programme, policy makers from around the world acknowledged that inequality in their countries is generally high and potentially a threat to long-term social and economic development.
- Evidence from developing countries shows that children in the poorest 20% of the populations are still up to three times more likely to die before their fifth birthday than children in the richest quintiles.
- Social protection has been significantly extended globally, yet persons with disabilities are up to five times more likely than average to incur catastrophic health expenditures.
- Despite overall declines in maternal mortality in the majority of developing countries, women in rural areas are still up to three times more likely to die while giving birth than women living in urban centres.

Goal 11: Sustainable Cities and Communities

Make Cities Inclusive, Safe, Resilient and Sustainable

Cities are hubs for ideas, commerce, culture, science, productivity, social development and much more. At their best, cities have enabled people to advance socially and economically.

However, many challenges exist to maintaining cities in a way that continues to create jobs and prosperity while not straining land and resources. Common urban challenges include congestion, lack of funds to provide basic services, a shortage of adequate housing and declining infrastructure.

The challenges cities face can be overcome in ways that allow them to continue to thrive and grow, while improving resource use and reducing pollution and poverty. The future we want includes cities of opportunities for all, with access to basic services, energy, housing, transportation and more.

Goal 11 Targets
- By 2030, ensure access for all to adequate, safe and affordable housing and basic services and upgrade slums.
- By 2030, provide access to safe, affordable, accessible and sustainable transport systems for all, improving road safety, notably by expanding public transport, with special attention to the needs of those in vulnerable situations, women, children, persons with disabilities and older persons.
- By 2030, enhance inclusive and sustainable urbanization and capacity for participatory, integrated and sustainable human settlement planning and management in all countries.

- Strengthen efforts to protect and safeguard the world's cultural and natural heritage.
- By 2030, significantly reduce the number of deaths and the number of people affected and substantially decrease the direct economic losses relative to global gross domestic product caused by disasters, including water-related disasters, with a focus on protecting the poor and people in vulnerable situations.
- By 2030, reduce the adverse per capita environmental impact of cities, including by paying special attention to air quality and municipal and other waste management.
- By 2030, provide universal access to safe, inclusive and accessible, green and public spaces, in particular for women and children, older persons and persons with disabilities.
- Support positive economic, social and environmental links between urban, peri-urban and rural areas by strengthening national and regional development planning.
- By 2020, substantially increase the number of cities and human settlements adopting and implementing integrated policies and plans towards inclusion, resource efficiency, mitigation and adaptation to climate change, resilience to disasters, and develop and implement, in line with the Sendai Framework for Disaster Risk Reduction 2015–2030, holistic disaster risk management at all levels.
- Support least developed countries, including through financial and technical assistance, in building sustainable and resilient buildings utilizing local materials.

Facts and Figures
- Half of humanity – 3.5 billion people – lives in cities today
- By 2030, almost 60% of the world's population will live in urban areas.
- 95% of urban expansion in the next decades will take place in developing world.
- 828 million people live in slums today and the number keeps rising.
- The world's cities occupy just 3% of the Earth's land, but account for 60–80% of energy consumption and 75% of carbon emissions.
- Rapid urbanization is exerting pressure on fresh water supplies, sewage, the living environment, and public health.
- But the high density of cities can bring efficiency gains and technological innovation while reducing resource and energy consumption.

Goal 12: Responsible Consumption and Production

Ensure Sustainable Consumption and Production Patterns

Sustainable consumption and production is about promoting resource and energy efficiency, sustainable infrastructure, and providing access to basic services, green and decent jobs and a better quality of life for all. Its implementation helps to achieve overall development plans, reduce future economic, environmental and social costs, strengthen economic competitiveness and reduce poverty.

Sustainable consumption and production aims at "doing more and better with less," increasing net welfare gains from economic activities by reducing resource use, degradation and pollution along the whole lifecycle, while increasing quality of life. It involves different stakeholders, including business, consumers, policy makers, researchers, scientists, retailers, media, and development cooperation agencies, among others.

It also requires a systemic approach and cooperation among actors operating in the supply chain, from producer to final consumer. It involves engaging consumers through awareness-raising and education on sustainable consumption and lifestyles, providing consumers with adequate information through standards and labels and engaging in sustainable public procurement, among others.

Goal 12 Targets
- Implement the 10-year framework of programmes on sustainable consumption and production, all countries taking action, with developed countries taking the lead, taking into account the development and capabilities of developing countries.
- By 2030, achieve the sustainable management and efficient use of natural resources.
- By 2030, halve per capita global food waste at the retail and consumer levels and reduce food losses along production and supply chains, including post-harvest losses.
- By 2020, achieve the environmentally sound management of chemicals and all wastes throughout their life cycle, in accordance with agreed international frameworks, and significantly reduce their release to air, water and soil in order to minimize their adverse impacts on human health and the environment.
- By 2030, substantially reduce waste generation through prevention, reduction, recycling and reuse.
- Encourage companies, especially large and transnational companies, to adopt sustainable practices and to integrate sustainability information into their reporting cycle.
- Promote public procurement practices that are sustainable, in accordance with national policies and priorities.
- By 2030, ensure that people everywhere have the relevant information and awareness for sustainable development and lifestyles in harmony with nature.
- Support developing countries to strengthen their scientific and technological capacity to move towards more sustainable patterns of consumption and production.
- Develop and implement tools to monitor sustainable development impacts for sustainable tourism that creates jobs and promotes local culture and products.
- Rationalize inefficient fossil-fuel subsidies that encourage wasteful consumption by removing market distortions, in accordance with national circumstances, including by restructuring taxation and phasing out those harmful subsidies, where they exist, to reflect their environmental impacts, taking fully into account

the specific needs and conditions of developing countries and minimizing the possible adverse impacts on their development in a manner that protects the poor and the affected communities

Facts and Figures

- Each year, an estimated one third of all food produced – equivalent to 1.3 billion tonnes worth around $1 trillion – ends up rotting in the bins of consumers and retailers, or spoiling due to poor transportation and harvesting practices.
- If people worldwide switched to energy efficient lightbulbs the world would save US$120 billion annually.
- Should the global population reach 9.6 billion by 2050, the equivalent of almost three planets could be required to provide the natural resources needed to sustain current lifestyles.
- Less than 3% of the world's water is fresh (drinkable), of which 2.5% is frozen in the Antarctica, Arctic and glaciers. Humanity must therefore rely on 0.5% for all of man's ecosystem's and fresh water needs.
- Man is polluting water faster than nature can recycle and purify water in rivers and lakes.
- More than one billion people still do not have access to fresh water.
- Excessive use of water contributes to the global water stress.
- Water is free from nature but the infrastructure needed to deliver it is expensive.
- Despite technological advances that have promoted energy efficiency gains, energy use in OECD countries will continue to grow another 35% by 2020. Commercial and residential energy use is the second most rapidly growing area of global energy use after transport.
- In 2002 the motor vehicle stock in OECD countries was 550 million vehicles (75% of which were personal cars). A 32% increase in vehicle ownership is expected by 2020. At the same time, motor vehicle kilometres are projected to increase by 40% and global air travel is projected to triple in the same period.
- Households consume 29% of global energy and consequently contribute to 21% of resultant CO_2 emissions.
- One-fifth of the world's final energy consumption in 2013 was from renewables.
- While substantial environmental impacts from food occur in the production phase (agriculture, food processing), households influence these impacts through their dietary choices and habits. This consequently affects the environment through food-related energy consumption and waste generation.
- 1.3 billion tonnes of food is wasted every year while almost one billion people go undernourished and another one billion hungry.
- Overconsumption of food is detrimental to our health and the environment.
- Two billion people globally are overweight or obese.
- Land degradation, declining soil fertility, unsustainable water use, overfishing and marine environment degradation are all lessening the ability of the natural resource base to supply food.
- The food sector accounts for around 30% of the world's total energy consumption and accounts for around 22% of total Greenhouse Gas emissions.

Goal 13: Climate Action

Take Urgent Action to Combat Climate Change and Its Impacts

Climate change is now affecting every country on every continent. It is disrupting national economies and affecting lives, costing people, communities and countries dearly today and even more tomorrow.

People are experiencing the significant impacts of climate change, which include changing weather patterns, rising sea level, and more extreme weather events. The greenhouse gas emissions from human activities are driving climate change and continue to rise. They are now at their highest levels in history. Without action, the world's average surface temperature is projected to rise over the twenty-first century and is likely to surpass 3 °C this century—with some areas of the world expected to warm even more. The poorest and most vulnerable people are being affected the most.

Affordable, scalable solutions are now available to enable countries to leapfrog to cleaner, more resilient economies. The pace of change is quickening as more people are turning to renewable energy and a range of other measures that will reduce emissions and increase adaptation efforts.

But climate change is a global challenge that does not respect national borders. Emissions anywhere affect people every.

Goal 13 Targets

- Strengthen resilience and adaptive capacity to climate-related hazards and natural disasters in all countries.
- Integrate climate change measures into national policies, strategies and planning.
- Improve education, awareness-raising and human and institutional capacity on climate change mitigation, adaptation, impact reduction and early warning.
- Implement the commitment undertaken by developed-country parties to the United Nations Framework Convention on Climate Change to a goal of mobilizing jointly $100 billion annually by 2020 from all sources to address the needs of developing countries in the context of meaningful mitigation actions and transparency on implementation and fully operationalize the Green Climate Fund through its capitalization as soon as possible.
- Promote mechanisms for raising capacity for effective climate change-related planning and management in least developed countries and small island developing States, including focusing on women, youth and local and marginalized communities.
- Acknowledging that the United Nations Framework Convention on Climate Change is the primary international, intergovernmental forum for negotiating the global response to climate change.

Facts and Figures

- From 1880 to 2012, average global temperature increased by 0.85 °C. To put this into perspective, for each 1 ° of temperature increase, grain yields decline by about 5%. Maize, wheat and other major crops have experienced significant yield reductions at the global level of 40 megatonnes per year between 1981 and 2002 due to a warmer climate.
- Oceans have warmed, the amounts of snow and ice have diminished and sea level has risen. From 1901 to 2010, the global average sea level rose by 19 cm as oceans expanded due to warming and ice melted. The Arctic's sea ice extent has shrunk in every successive decade since 1979, with 1.07 million km^2 of ice loss every decade.
- Given current concentrations and on-going emissions of greenhouse gases, it is likely that by the end of this century, the increase in global temperature will exceed 1.5 °C compared to 1850–1900 for all but one scenario. The world's oceans will warm and ice melt will continue. Average sea level rise is predicted as 24–30 cm by 2065 and 40–63 cm by 2100. Most aspects of climate change will persist for many centuries even if emissions are stopped.
- Global emissions of carbon dioxide (CO_2) have increased by almost 50% since 1990.
- Emissions grew more quickly between 2000 and 2010 than in each of the three previous decades
- It is still possible, using a wide array of technological measures and changes in behaviour, to limit the increase in global mean temperature to two degrees Celsius above pre-industrial levels.
- Major institutional and technological change will give a better than even chance that global warming will not exceed this threshold.

Goal 14: Life Below Water

Conserve and Sustainably Use the Oceans, Seas and Marine Resources

The world's oceans – their temperature, chemistry, currents and life – drive global systems that make the Earth habitable for humankind.

Our rainwater, drinking water, weather, climate, coastlines, much of our food, and even the oxygen in the air we breathe, are all ultimately provided and regulated by the sea. Throughout history, oceans and seas have been vital conduits for trade and transportation.

Careful management of this essential global resource is a key feature of a sustainable future.

Goal 14 Targets
- By 2025, prevent and significantly reduce marine pollution of all kinds, in particular from land-based activities, including marine debris and nutrient pollution.
- By 2020, sustainably manage and protect marine and coastal ecosystems to avoid significant adverse impacts, including by strengthening their resilience, and take action for their restoration in order to achieve healthy and productive oceans.
- Minimize and address the impacts of ocean acidification, including through enhanced scientific cooperation at all levels. By 2020, effectively regulate harvesting and end overfishing, illegal, unreported and unregulated fishing and destructive fishing practices and implement science-based management plans, in order to restore fish stocks in the shortest time feasible, at least to levels that can produce maximum sustainable yield as determined by their biological characteristics.
- By 2020, conserve at least 10% of coastal and marine areas, consistent with national and international law and based on the best available scientific information.
- By 2020, prohibit certain forms of fisheries subsidies which contribute to overcapacity and overfishing, eliminate subsidies that contribute to illegal, unreported and unregulated fishing and refrain from introducing new such subsidies, recognizing that appropriate and effective special and differential treatment for developing and least developed countries should be an integral part of the World Trade Organization fisheries subsidies negotiation.
- By 2030, increase the economic benefits to Small Island developing States and least developed countries from the sustainable use of marine resources, including through sustainable management of fisheries, aquaculture and tourism.
- Increase scientific knowledge, develop research capacity and transfer marine technology, taking into account the Intergovernmental Oceanographic Commission Criteria and Guidelines on the Transfer of Marine Technology, in order to improve ocean health and to enhance the contribution of marine biodiversity to the development of developing countries, in particular small island developing States and least developed countries.
- Provide access for small-scale artisanal fishers to marine resources and markets.
- Enhance the conservation and sustainable use of oceans and their resources by implementing international law as reflected in UNCLOS, which provides the legal framework for the conservation and sustainable use of oceans and their resources, as recalled in paragraph 158 of The Future We Want.

Facts and Figures
- Oceans cover three quarters of the Earth's surface, contain 97 per cent of the Earth's water, and represent 99% of the living space on the planet by volume.
- Over three billion people depend on marine and coastal biodiversity for their livelihoods.

- Globally, the market value of marine and coastal resources and industries is estimated at $3 trillion per year or about 5% of global GDP.
- Oceans contain nearly 200,000 identified species, but actual numbers may lie in the millions.
- Oceans absorb about 30% of carbon dioxide produced by humans, buffering the impacts of global warming.
- Oceans serve as the world's largest source of protein, with more than three billion people depending on the oceans as their primary source of protein.
- Marine fisheries directly or indirectly employ over 200 million people.
- Subsidies for fishing are contributing to the rapid depletion of many fish species and are preventing efforts to save and restore global fisheries and related jobs, causing ocean fisheries to generate US$ 50 billion less per year than they could.
- As much as 40% of the world oceans are heavily affected by human activities, including pollution, depleted fisheries, and loss of coastal habitats.

Goal 15: Life on Land

Sustainably Manage Forests, Combat Desertification, Halt and Reverse Land Degradation, Halt Biodiversity Loss

Forests cover 30% of the Earth's surface and in addition to providing food security and shelter, forests are key to combating climate change, protecting biodiversity and the homes of the indigenous population. Thirteen million hectares of forests are being lost every year while the persistent degradation of drylands has led to the desertification of 3.6 billion hectares.

Deforestation and desertification – caused by human activities and climate change – pose major challenges to sustainable development and have affected the lives and livelihoods of millions of people in the fight against poverty. Efforts are being made to manage forests and combat desertification.

Goal 15 Targets
- By 2020, ensure the conservation, restoration and sustainable use of terrestrial and inland freshwater ecosystems and their services, in particular forests, wetlands, mountains and drylands, in line with obligations under international agreements.
- By 2020, promote the implementation of sustainable management of all types of forests, halt deforestation, restore degraded forests and substantially increase afforestation and reforestation globally.
- By 2030, combat desertification, restore degraded land and soil, including land affected by desertification, drought and floods, and strive to achieve a land degradation-neutral world.

- By 2030, ensure the conservation of mountain ecosystems, including their biodiversity, in order to enhance their capacity to provide benefits that are essential for sustainable development.
- Take urgent and significant action to reduce the degradation of natural habitats, halt the loss of biodiversity and, by 2020, protect and prevent the extinction of threatened species.
- Promote fair and equitable sharing of the benefits arising from the utilization of genetic resources and promote appropriate access to such resources, as internationally agreed.
- Take urgent action to end poaching and trafficking of protected species of flora and fauna and address both demand and supply of illegal wildlife products.
- By 2020, introduce measures to prevent the introduction and significantly reduce the impact of invasive alien species on land and water ecosystems and control or eradicate the priority species.
- By 2020, integrate ecosystem and biodiversity values into national and local planning, development processes, poverty reduction strategies and accounts.
- Mobilize and significantly increase financial resources from all sources to conserve and sustainably use biodiversity and ecosystems.
- Mobilize significant resources from all sources and at all levels to finance sustainable forest management and provide adequate incentives to developing countries to advance such management, including for conservation and reforestation.
- Enhance global support for efforts to combat poaching and trafficking of protected species, including by increasing the capacity of local communities to pursue sustainable livelihood opportunities.

Facts and Figures
- Around 1.6 billion people depend on forests for their livelihood. This includes some 70 million indigenous people.
- Forests are home to more than 80% of all terrestrial species of animals, plants and insects.
- 2.6 billion people depend directly on agriculture, but 52% of the land used for agriculture is moderately or severely affected by soil degradation.
- As of 2008, land degradation affected 1.5 billion people globally.
- Arable land loss is estimated at 30–35 times the historical rate.
- Due to drought and desertification each year 12 million hectares are lost (23 hectares per minute), where 20 million tons of grain could have been grown.
- 74% of the poor are directly affected by land degradation globally.
- Of the 8300 animal breeds known, 8% are extinct and 22% are at risk of extinction.
- Of the over 80,000 tree species, less than 1% have been studied for potential use.
- Fish provide 20% of animal protein to about three billion people. Only ten species provide about 30% of marine capture fisheries and ten species provide about 50% of aquaculture production.

- Over 80% of the human diet is provided by plants. Only three cereal crops – rice, maize and wheat – provide 60% of energy intake.
- As many as 80% of people living in rural areas in developing countries rely on traditional plant-based medicines for basic.
- Micro-organisms and invertebrates are key to ecosystem services, but their contributions are still poorly known and rarely acknowledged.

Goal 16: Peace, Justice and Strong Institutions

Promote Just, Peaceful and Inclusive Societies

Goal 16 of the Sustainable Development Goals is dedicated to the promotion of peaceful and inclusive societies for sustainable development, the provision of access to justice for all, and building effective, accountable institutions at all levels.

Goal 16 Targets
- Significantly reduce all forms of violence and related death rates everywhere.
- End abuse, exploitation, trafficking and all forms of violence against and torture of children.
- Promote the rule of law at the national and international levels and ensure equal access to justice for all.
- By 2030, significantly reduce illicit financial and arms flows, strengthen the recovery and return of stolen assets and combat all forms of organized crime.
- Substantially reduce corruption and bribery in all their forms.
- Develop effective, accountable and transparent institutions at all levels.
- Ensure responsive, inclusive, participatory and representative decision-making at all levels.
- Broaden and strengthen the participation of developing countries in the institutions of global governance.
- By 2030, provide legal identity for all, including birth registration.
- Ensure public access to information and protect fundamental freedoms, in accordance with national legislation and international agreements.
- Strengthen relevant national institutions, including through international cooperation, for building capacity at all levels, in particular in developing countries, to prevent violence and combat terrorism and crime.
- Promote and enforce non-discriminatory laws and policies for sustainable development.

Facts and Figures
- Among the institutions most affected by corruption are the judiciary and police.
- Corruption, bribery, theft and tax evasion cost some US $1.26 trillion for developing countries per year; this amount of money could be used to lift those who are living on less than $1.25 a day above $1.25 for at least 6 years.

- The rate of children leaving primary school in conflict affected countries reached 50% in 2011, which accounts to 28.5 million children, showing the impact of unstable societies on one of the major goals of the post 2015 agenda: education.
- The rule of law and development have a significant interrelation and are mutually reinforcing, making it essential for sustainable development at the national and international level.

Goal 17: Partnerships for the Goals

Revitalize the Global Partnership for Sustainable Development

A successful sustainable development agenda requires partnerships between governments, the private sector and civil society. These inclusive partnerships built upon principles and values, a shared vision, and shared goals that place people and the planet at the centre, are needed at the global, regional, national and local level.

Urgent action is needed to mobilize, redirect and unlock the transformative power of trillions of dollars of private resources to deliver on sustainable development objectives. Long-term investments, including foreign direct investment, are needed in critical sectors, especially in developing countries. These include sustainable energy, infrastructure and transport, as well as information and communications technologies. The public sector will need to set a clear direction. Review and monitoring frameworks, regulations and incentive structures that enable such investments must be retooled to attract investments and reinforce sustainable development. National oversight mechanisms such as supreme audit institutions and oversight functions by legislatures should be strengthened.

Goals 17 Targets
- Strengthen domestic resource mobilization, including through international support to developing countries, to improve domestic capacity for tax and other revenue collection.
- Developed countries to implement fully their official development assistance commitments, including the commitment by many developed countries to achieve the target of 0.7% of ODA/GNI to developing countries and 0.15–0.20% of ODA/GNI to least developed countries ODA providers are encouraged to consider setting a target to provide at least 0.20% of ODA/GNI to least developed countries. Mobilize additional financial resources for developing countries from multiple sources.
- Assist developing countries in attaining long-term debt sustainability through coordinated policies aimed at fostering debt financing, debt relief and debt restructuring, as appropriate, and address the external debt of highly indebted poor countries to reduce debt distress.
- Adopt and implement investment promotion regimes for least developed countries.

- Enhance North-South, South-South and triangular regional and international cooperation on and access to science, technology and innovation and enhance knowledge sharing on mutually agreed terms, including through improved coordination among existing mechanisms, in particular at the United Nations level, and through a global technology facilitation mechanism.
- Promote the development, transfer, dissemination and diffusion of environmentally sound technologies to developing countries on favourable terms, including on concessional and preferential terms, as mutually agreed.
- Fully operationalize the technology bank and science, technology and innovation capacity-building mechanism for least developed countries by 2017 and enhance the use of enabling technology, in particular information and communications technology.
- Enhance international support for implementing effective and targeted capacity-building in developing countries to support national plans to implement all the sustainable development goals, including through North-South, South-South and triangular cooperation.
- Promote a universal, rules-based, open, non-discriminatory and equitable multilateral trading system under the World Trade Organization, including through the conclusion of negotiations under its Doha Development Agenda.
- Significantly increase the exports of developing countries, in particular with a view to doubling the least developed countries' share of global exports by 2020.
- Realize timely implementation of duty-free and quota-free market access on a lasting basis for all least developed countries, consistent with World Trade Organization decisions, including by ensuring that preferential rules of origin applicable to imports from least developed countries are transparent and simple, and contribute to facilitating market access.
- Enhance global macroeconomic stability, including through policy coordination and policy coherence.
- Enhance policy coherence for sustainable development.
- Respect each country's policy space and leadership to establish and implement policies for poverty eradication and sustainable development.
- Enhance the global partnership for sustainable development, complemented by multi-stakeholder partnerships that mobilize and share knowledge, expertise, technology and financial resources, to support the achievement of the sustainable development goals in all countries, in particular developing countries.
- Encourage and promote effective public, public-private and civil society partnerships, building on the experience and resourcing strategies of partnerships.
- By 2020, enhance capacity-building support to developing countries, including for least developed countries and small island developing States, to increase significantly the availability of high-quality, timely and reliable data disaggregated by income, gender, age, race, ethnicity, migratory status, disability, geographic location and other characteristics relevant in national contexts.
- By 2030, build on existing initiatives to develop measurements of progress on sustainable development that complement gross domestic product, and support statistical capacity-building in developing countries

Facts and Figures

- Official development assistance stood at $135.2 billion in 2014, the highest level ever recorded.
- 79% of imports from developing countries enter developed countries duty-free.
- The debt burden on developing countries remains stable at about 3% of export revenue.
- The number of Internet users in Africa almost doubled in the past 4 years.
- 30% of the world's youth are digital natives, active online for at least 5 years.
- But four billion more people do not use the Internet, and 90% of them are from the developing world

Appendix 2: Key Issues of the European General Data Protection Regulation (GDPR)

Source: https://gdpr-info.eu/key-issues/ [1]

Key Issues

Welcome to the section "Key Issues". Under the various keywords you can find a brief introduction and the Articles of the GDPR as well as the recitals that are relevant to the topic.

For more detailed information we compiled a list of links with expert contributions and opinions of the data protection authorities which are provided at https://gdpr-info.eu/key-issues/. The information provided at these links is only a possible interpretation of the law which is not legally binding. The final interpretation of the GDPR is exclusively within the jurisdiction of the European Court of Justice. However, the opinions of the supervisory authorities are of considerable practical relevance due to their supervision through their investigative and corrective powers.

---

[1] **Author's Note**: The full text of the European General Data Protection Regulation (GDPR) that includes 11 chapters and 99 articles can be found at https://gdpr-info.eu. The complete provisions of the GDPR as adopted by the European Union entered into effect as of May 25, 2018. Every large corporation, organization, or agency, now operating on a global basis – and thus also operating in Europe – are now impacted by these provisions. The GDPR may in time provide a legal basis and key precedent for what other countries do with respect data protection for individuals and various entities around the world. Since these provisions are so comprehensive and potentially impactful on all nations of the world the following 14 key issues of the GDPR as outlined and summarized by in the EU "key issues" presentation are provided below. The recitals and reference links in the Key Issues presentations are not included. This information, however, can be found at: https://gdpr-info.eu/key-issues/

© Springer Nature Switzerland AG 2019   201
J. N. Pelton, *Preparing for the Next Cyber Revolution*,
https://doi.org/10.1007/978-3-030-02137-5

Consent

Processing personal data is generally forbidden if it is not expressly allowed by law, or the impacted persons have not consented to processing these data. The consent of those whose personal data is collected, processed and/or used puts the persons in the position to be able to dispose of their personal rights.

The basic requirements for the effectiveness of valid legal consent are defined in Art. 7 of the GDPR and specified further in recital 32. This must be voluntarily granted for a concrete case after sufficient information is provided to the person involved and must be clearly communicated. The person involved must be given a true choice for the consent to be voluntary. In addition, a so-called "coupling prohibition" applies. Thus, a concluded contract may not be made dependent upon the consent to process further personal data which is not needed for completing a transaction. In addition, the consent must be bound to one or several specified purposes which are then sufficiently explained. Should the consent legitimise the processing of special personal data, it must expressly refer to this. The person impacted must, in all cases, be explained the ability to retract his consent. The retraction must be as easy to do as the granting of the consent itself.

There is no form requirement for the consent, even if a written consent is recommended due to the accountability of the persons responsible. It can therefore be executed in electronic form. One must consider, however, that recital 32 requires that a consent may only be granted through a clear negotiation. This includes the requirement for an opt-in. A special item in this regard is consent for children and adolescents in relation to information company services. There is an additional consent or agreement requirement from those with parental rights for those who are under the age of 16. The age limit is subject to an escape clause. Member States can reduce this in their national law to 13 years of age. Should the service offering not explicitly be directed to children, it is freed of this rule. This does not apply, however, to offers which are open to both children and adults.

Suitable GDPR Articles

Art. 4 GDPR Definitions Art. 6 GDPR Lawfulness of processing Art. 7 GDPR Conditions for consent Art. 8 GDPR Conditions applicable to child's consent in relation to information society services Art. 9 GDPR Processing of special categories of personal data Art. 22 GDPR Automated individual decision-making, including profiling Art. 49 GDPR Derogations for specific situations.

Data Protection Officer

The concept of a Data Protection Officer was created in Europe with the basic data protection regulation. The obligation to appoint a Data Protection Officer impacts companies depending upon their core activities which are essential to achieving their goals. This includes processing of personal data or data processing which is particularly decisive to those impacted. These companies must appoint an operational Data Protection Officer. In addition, the regulation to appoint a Data Protection Officer also has an escape clause for Member States. These States are free to decide whether to appoint an operating Data Protection Officer at a company under narrow conditions. If such an obligation exists under the basic data protection regulation or a national law, corporate groups can also appoint a joint operating Data Protection Officer. Their location must be easily reached by external stakeholders, supervisory authorities and employees.

Groups and companies have two possibilities to meet their obligation to appoint a Data Protection Officer. Either they name an employee as an internal Data Protection Officer, or they appoint an external Data Protection Officer. In selecting such a person, they must ensure that an internal Data Protection Officer is not subject to conflicts of interest, such as because he is an employee in the IT Department, HR Department or senior management, and must inspect himself. Regardless of which option is chosen, a Data Protection Officer must provide some professional knowledge in data protection law and IT security which includes the complexity of data processing and the size of the company.

Duties of the Data Protection Officer include: Acting on the compliance to all relevant data protection regulations, monitoring specific processes, such as data protection impact assessments, employee awareness and training employees, as well as collaboration with authorities. Therefore, the operating Data Protection Officer must not be recalled or disadvantaged due to his fulfillment of his tasks. Despite the monitoring function, the company itself remains responsible for compliance with data protection regulations. The Data Protection Officer is therefore bound to "properly and in a timely manner, in all issues which relate to the protection of personal data". When the Data Protection Officer is appointed, his superior must publish his contact data, and communicate his appointment and contact data to authorities.

Willful or negligent failure to appoint a corporate Data Protection Officer is an offence subject to fines.

Suitable GDPR Articles
Art. 35 GDPR Data protection impact assessment Art. 37 GDPR Designation of the data protection officer Art. 38 GDPR Position of the data protection officer Art. 39 GDPR Tasks of the data protection officer.

Email Marketing

Newsletter mailing, and email marketing are part of the fixed online marketing universe. Basically, prohibition with opt-in permission also applies here for the processing of personal data. Processing is only allowed by law if either there is a consent by those impacted, or a statutory justification. This could be, for example, preserving the justified interests of the person responsible for sending email marketing. Recital 47 of the General Data Protection Regulation expressly states that the law also applies to the processing of personal data for direct marketing as a legitimate interest of the person responsible.

In addition, such an interest could be present, for example, if there is a relevant and proportionate relationship between the persons concerned and the person responsible. This could be the case if the person involved is a customer of the person responsible or is in the latter's service. Therefore, much indicates that email marketing is allowed without consent, at least for existing customers. If the company has a justified interest in 'cold' acquisition through email marketing, the marketing emails can be allowed to potential customers without consent. To receive no further information by newsletter or email, the customer receiving them need only object to processing for marketing purposes.

One must note, however, that according to Art. 95 of the General Data Protection Regulation, this applies to all data protection-related purposes unless special rules with the same regulatory target are contained in the ePrivacy directive (see also recital 173). The consequence is that email marketing is currently only allowed with the consent of those impacted (Art. 13, para. 1 of RL 2002/58/EC). One must wait to see whether the coming ePrivacy regulation provides more clarity about this issue.

Regardless of whether a company supports the marketing measures afterwards on its justified interest or on consent, the obligation arises for the person responsible to adhere to comprehensive information obligations. Their contents are differentiated using which justification reason is selected.

Suitable GDPR Articles
Art. 6 GDPR Lawfulness of processing Art. 7 GDPR Conditions for consent Art. 21 GDPR Right to object Art. 95 GDPR Relationship with Directive 2002/58/EC.

Encryption

Using encryption of personal data, companies can reduce the probability of a data failure, and therefore also fines, in the future. The processing of personal data is naturally associated with a certain degree of risk. Especially these days, cyberattacks are nearly unavoidable for companies above a given size. Therefore, risk management plays an ever-larger role in IT security. Data encryption is therefore suited, among other means, for these companies.

In general, one understands encryption as a procedure that converts clear text into a hashed code using a key, as that the outgoing information can only become readable again by using the correct key. This minimizes the risk of an incident during data processing, as encrypted contents are basically unreadable for third parties who don't have the correct key. Encryption is the best means to protect data during transfer, and a way to secure stored personal data. This reduces the abuse risk within a company, as access is limited only to authorized people with the right key.

The regulation also recognizes risks when processing personal data and places the responsibility on the responsible parties in Art. 32 (1) of the General Data Protection Regulation to use suitable technical and organizational measures to secure personal data. One must consider the state of the art, implementation costs and the type, scope, circumstances and purpose of the processing. In addition to these criteria, various access probabilities and the severity of the risks to the rights and freedoms of those impacted must be considered. One must adjust the degree of the security measures taken because of the above consideration. Encryption is therefore explicitly mentioned as such a measure in the list of Art. 32(1) of the GDPR, which is not exhaustive.

Encryption of personal data has additional benefits for responsible parties and/or order processors. So, if one loses a mobile medium on which data are encrypted using state of the art methods need not be reported, as a rule. In addition, the authorities must positively consider the use of encryption in their decision on whether and to what level a sanction is assessed as per Art. 83, para. 2 let. C of the GDPR.

Suitable GDPR Articles
Art. 6 GDPR Lawfulness of processing Art. 32 GDPR Security of processing Art. 34 GDPR Communication of a personal data breach to the data subject.

Fines/Penalties

National authorities can or must assess fines for specific data protection violations in accordance with the General Data Protection Regulation. The fines are applied additionally or instead of further remedies or powers, such as the order to end a violation, an instruction to adjust the data processing to statutory requirements, as well as the granting of a prohibition which is limited in time or permanently, to perform data processing. For the provisions which relate to order processors, they can be directly and/or in conjunction with the person responsible, subject to sanctions.

The fines must be effective, reasonable and dissuasive for each individual case. For the decision of whether and what amount of sanctions can be assessed, the authorities have a statutory catalogue of criteria which must be used in taking a decision. Among other things, intentional infringement, a failure to take measures to mitigate the damage which occurred, or lack of collaboration with authorities can increase the penalties. For the especially severe violations listed in Art. 83, para. 5 of the GDPR, the fine framework can be up to 20 million euros, or in the case of a company, up to 4% of their total global turnover in the previous fiscal year, whichever is higher. But even the catalogue of less severe violations (Art. 83, para. 4) sets forth fines of up to 10,000,000 euros, or, in the case of a company, up to 2% of its entire global turnover of the previous fiscal year, whichever is higher. Especially important here is that the term "company" is equivalent to that used in Art. 101 and 102 of the Treaty on the Functioning of the European Union (TFEU). According to case law from the Court of Justice of the European Union, this refers to the broad, functional corporate term as a company which is a unit which exercises a commercial activity, independent of its legal form and its type of financing. This commercial unit can therefore consist of one individual company in the sense of a legal subject, but out of several natural or legal persons. Thus, a whole group can be treated as one company. To calculate fines, the entire group's turnover is used to calculate a penalty based on the company's turnover. In addition, Member States have rules for sanctions for other violations against the Regulation. This applies to those violations to which a fine has not already been assessed. Therefore, one must ensure that these penalties are also effective, proportional and act as a deterrent.

An objectionable fact in the company can be found through proactive inspection activities conducted by the assigned authorities, by an unsatisfied employee who complains to the authorities or by customers or potential customers who register a notice to the authorities, through the company making its own declaration, or by the press in general, through investigative journalism.

Suitable GDPR Articles
Art. 58 GDPR Powers Art. 70 GDPR Tasks of the Board Art. 83 GDPR General conditions for imposing administrative fines Art. 84 GDPR Penalties.

Information Obligations

One requires transparency in gathering and using data in order to allow EU citizens to exercise their rights to personal data. Therefore, the General Data Protection Regulation sets forth a variety of information obligations.

The law differentiates between two cases: On the one hand, if personal data is directly obtained from the impacted party (Art. 13 of the GDPR) and, on the other hand, if this is not directly obtained from the impacted person (Art. 14 of the GDPR). For direct obtaining of information, the person must be immediately informed.

In terms of content, the obligation to inform of the responsible party include identity, contact data of the Data Protection Officer (if available), the processing purposes and the legal bases, any justified interest, about the receiver when transmitting the data, and also about any transfer to third countries. In addition, the information obligation also includes information about the duration of storage, the rights of the impacted parties, the ability to withdraw consents, the right to complain to the authorities, as well as the statutory or contractual obligation to provide personal data. In addition, they must be informed of any automated decision-taking or other profiling activities. This information requirement can be dispensed for direct collection only if the impacted person already has this information.

If the information gathering is not done with the impacted person, this person must be informed within a reasonable period of time, but at latest after a month, when using this information for communication, inform them through direct contact. As far as content is concerned, the responsible party is also subject to the same information obligations with this type of information gathering. The only exception is only information about the obligation to provide, as the responsible party cannot decide about this on his own. In addition, he has the obligation to inform from what sources the data originated, and whether it was publicly available. The information obligations must be provided in a precise, transparent, comprehensible and easily accessible form. This can be communicated to the impacted person in writing or electronic form. It is explicitly explained that also so-called 'standardised image symbols' can be used in order to convey a meaningful overview of the intended processing in an easily comprehended, understandable and clear form.

In the case that the personal data is not gathered from the impacted party, the information obligation need not be fulfilled in exceptional cases. This applies if this is either impossible or unreasonably expensive, the gathering and/or transmission is required by law, or if professional secrecy or other statutory secrecy obligation is in place.

Suitable GDPR Articles

Art. 12 GDPR Transparent information, communication and modalities for the exercise of the rights of the data subject Art. 13 GDPR Information to be provided where personal data are collected from the data subject Art. 14 GDPR Information to be provided where personal data have not been obtained from the data subject.

Order Processing

The General Data Protection Regulation offers a uniform, Europe-wide ability for so-called 'order processing'. Order processing is the gathering, processing or use of personal data by an order processor in accordance with the instructions of those responsible for the data processing based on a contract.

The relevant regulations for order processing are already applied if the processing is connected to activities of a branch within the EU. This means that it is sufficient if either the person responsible or the order processor operates a branch in the EU, and the processing is connected to this work. In a constellation of order processing, the joint persons responsible (Art. 26 of the GDPR) are those who together define the purposes and means for the data processing, and who are also jointly responsible for these. The persons responsible must ensure, in selecting the order processor, that it has implemented sufficient technical and organizational measures to ensure that the rules of the regulation are complied with.

In most cases, order processing proceeds based on a contract. Art. 28, para. 3 of the GDPR sets forth the minimum requirements. This must contain, among other things, what type of personal data will be processed, as well as the object and purpose of the processing. In addition, there are further obligations for the order processor. For example, it must also maintain an index of the processing activities which includes the names and contact data for each person responsible who is working on which order, as well as the processing categories which are conducted for them. Furthermore, the index must include, if applicable, the transfer of personal data to third countries and, if possible, a general description of technical and organisational measures.

Basically, the person responsible is the first contact for those impacted, and for compliance with data processing legal requirements. This does not mean, however, that the order processor is free of liability. According to Art. 82 of the GDPR, he is jointly liable with the persons responsible. However, the order processor's liability is limited as per para. 2 to violations of duties which are specific to him. Both parties have the ability to exculpate themselves. To do this, they must prove that they were not responsible in any way for the circumstances leading to the damages.

Suitable GDPR Articles

Art. 4 GDPR Definitions Art. 27 GDPR Representatives of controllers or processors not established in the Union Art. 28 GDPR Processor Art. 29 GDPR Processing under the authority of the controller or processor Art. 30 GDPR Records of processing activities Art. 40 GDPR Codes of conduct Art. 42 GDPR Certification Art. 44 GDPR General principle for transfers Art. 45 GDPR Transfers on the basis of an adequacy decision Art. 46 GDPR Transfers subject to appropriate safeguards Art. 47 GDPR Binding corporate rules Art. 82 GDPR Right to compensation and liability.

Personal Data

The term 'personal data' is the entryway to application of the Data Protection Basic Regulation and is defined in Art. 4 para. 1 no. 1. Personal data are all information which is related to an identified or identifiable natural person.

Those impacted are identifiable if they can be identified, especially using assignment to an identifier such as a name, an identifying number, location data, an online identifier or one of several special characteristics, which expresses the physical, physiological, genetic, mental, commercial, cultural or social identity of these natural persons. In practice, these also include all data which are or can be assigned to a person in any kind of way. For example, the telephone number, credit card or personnel number of a person, account data, number plate, appearance and customer number or address are all personal data.

Since the definition includes "all information," one must assume that the term "personal data" should be as broadly interpreted as possible. This is also found in case law of the Court of Justice of the European Union. These include also less-clear information, such as recordings of work times which include information about the time when an employee begins and ends his work day, as well as breaks or times which do not fall in work time. Also, written answers from a test-taker and any remarks from the test about these answers are "personal data" if the test-taker can be theoretically identified. The same also applies to IP addresses. If the processor has the legal option to oblige the provider to publish additional information which can identify the user who is behind the IP address, this is also personal data. In addition, one must note that personal data need not be objective. Subjective information such as opinions, judgments or estimates can be personal data. Thus, this includes an assessment of creditworthiness of a person or an estimate of work performance by an employer.

Last but not least, the law states that the information for a personnel reference must refer to a natural person. In other words, data protection does not apply to information about legal entities such as corporations, foundations and institutions.

For natural persons, on the other hand, protection begins and is extinguished with legal capacity. Basically, a person obtains this capacity with his birth, and loses it upon his death. Data must therefore be assignable to specific or specifiable living persons for reference to a person.

In addition to general personal data, one must consider above all the special categories of personal data (also known as sensitive personal data) which are highly relevant because they are subject to a higher level of protection. These data include genetic, biometric and health data, as well as personal data from which racial and ethnic origin, political opinions, religious or ideological convictions or membership in a union can be attributed to a person.

Suitable GDPR Articles
Art. 4 GDPR Definitions Art. 9 GDPR Processing of special categories of personal data.

Privacy by Design

"Privacy by Design" and "Privacy by Default" have been frequently-discussed topics related to data protection. The first thoughts of "Privacy by Design" were expressed in the 1970s and were incorporated in the 1990s into the RL 95/46/EC data protection directive. According to recital 46 in this Directive, technical and organizational measures (TOM) must be taken already at the time of planning a processing system to protect data safety.

The term "Privacy by Design" means nothing more than "data protection through technology design." Behind this is the thought that data protection in data processing procedures is best adhered to when it is already integrated in the technology when created. Nevertheless, there is still uncertainty about what "Privacy by Design" means, and how one can implement it. This is due, on the one hand, to incomplete implementation of the Directive in some Member States and, on the other hand, that the principle "Privacy by Design" which is in the General Data Protection Regulation, that the current approach in the data protection guidelines, which requires persons responsible already to include definitions of the means for processing TOMs at the time that they are defined in order to fulfill the basics and requirements of "Privacy by Design". Legislation leaves completely open which exact protective measures are to be taken. As an example, one only need name pseudonymization. No more detail is given in recital 78 of the regulation. At least in other parts of the law, encryption is named, as well as anonymization of data as possible protective measures. Furthermore, user authentication and technical implementation of the right to object must be considered. In addition, when selecting precautions, one can use other standards, such as ISO standards. When selecting in individual cases, one must ensure that the state of the art as well as reasonable implementation costs are included.

In addition to the named criteria, the type, scope, circumstances and purpose of the processing must be considered. This must be contrasted with the various probability of occurrence and the severity of the risks connected to the processing. The text of the law leads one to conclude that often several protective measures must be used with one another to satisfy statutory requirements. In practice, this consideration is already performed in an early development phase when setting technology decisions. Recognised certification can serve as an indicator to authorities that the persons responsible have complied with the statutory requirements of "Privacy by Design".

Suitable GDPR Articles
Art. 25 GDPR Data protection by design and by default.

Privacy Impact Assessments

The instrument for a privacy impact assessment (PIA) or data protection impact assessment (DPIA) was introduced with the EU data protection basic regulation (Art. 35 of the GDPR). This has to do with the obligation of persons responsible to conduct an impact assessment and to document it before starting planned data processing. One can bundle the assessment for several processing procedures.

Basically, a data protection impact assessment must always be conducted when the processing could result in a high risk to the rights and freedoms of natural persons. In addition, it must be performed if one of the rule examples set forth in Art. 35(3) of the GDPR is relevant. The supervisory authorities should specify the formulation adhered to for the basic implementation obligation. In a first draft, the Article 29 Working Party created ten criteria which form an index for high risk to the rights and freedoms of a natural person, such as scoring/profiling, automatic decisions, which lead to legal consequences for those impacted, systematic monitoring, processing of special personal data, data which is processed in a large scope, the putting together or combining of data which was gathered by various processes, data about incapacitated people, or those with limited ability to act, use of newer technologies or biometric procedures, data transfer to countries outside the EU/EEC, and data processing which hinders those involved in exercising their rights. A privacy impact assessment is not necessary if a processing procedure fulfil only one of these criteria. If several criteria are met, however, the risk to those involved is higher, and a data protection impact assessment is absolutely required. If there is doubt, and it is difficult to find the limits, a DPIA must always be conducted. This must be repeated at least every 3 years.

In addition, authorities must be given a list in their area of responsibility and be published, and the procedural operations must be displayed in which a privacy impact assessment must be performed. They are also free to publish procedural operations must be published that do not specifically require publication. If a company has appointed a Data Protection Officer, his advice must be used when conducting the DPIA. How and by what criteria the impacts and risks for those impacted are assessed are open for the most part. The first templates were related to the inspection schemes of ISO standards or standard data protection models.

Suitable GDPR Articles
Art. 5 GDPR Principles relating to processing of personal data Art. 35 GDPR Data protection impact assessment Art. 36 GDPR Prior consultation Art. 57 GDPR Tasks.

Records of Processing Activities

The data protection basic regulation obligates, as per Art. 30 of the GDPR, written documentation and overview of procedures by which personal data are processed. Records of processing activities must include significant information about data processing, including data categories, the group of impacted people, the purpose of the processing and the data receivers. This must be completely provided to authorities upon request.

The obligation to create records of processing activities is not only incumbent upon persons responsible and their representative, but also directly on processing employees and their representatives as in Art. 30(2) of the GDPR. Companies or institutions with fewer than 250 employees are exceptionally freed from creating an index if the processing undertaken does not pose a risk to the rights and freedoms of those concerned, if no processing of special data categories is done, or if the processing is done only occasionally as it is indicated in Art. 30(5) GDPR. In practice, this waiver is rarely applicable. Apart from any difficulties which occur in designing what is "only occasional," most companies – even with a broad interpretation of the term – must clearly create regular data processing procedures, including for the website, their web shop, salary calculation or CRM systems. Above all, companies which have had no procedural index, will be subject to additional bureaucratic expense. Thus, one must note that the obligation for documentation and therefore records of processing activities will be a focus of authorities' inspections within the data protection basic regulation.

If a company does not maintain records of processing activities and/or does not provide a complete index to authorities, they are subject to fines according to Art. 83(4)(a) of the GDPR. The possible fines can be up to 10 million euros or 2% of their annual turnover. This total is, as a rule, only assessed by the authorities in exceptional cases. For this, the authorities must, as set forth in recital 13, "consider the special needs of the smallest companies as well as small and medium companies in the application of this regulation."

Suitable GDPR Articles
Art. 5 GDPR Principles relating to processing of personal data Art. 30 GDPR Records of processing activities.

Right of Access

The right of access plays a central role in the General Data Protection Regulation. On the one hand, as soon as the right of access becomes possible, further rights (such as authorization and erasure) must apply. On the other hand, information that is omitted or incomplete is subject to fine.

The answer to a request for information includes two stages. First, the responsible person must check whether any personal data of the person seeking information is being processed at all. In this case, one must report a positive or negative result. Should the answer be positive, the second stage includes a bandwidth of information. The right of access includes information about the processing purposes, the processing category of personal data, the receiver or categories of receivers, the planned duration of storage or criteria for their definition, information about the rights of those impacted such as correction, erasure or restrictions to processing, the right to object to this processing, instructions on the complaint rights to the authorities, information about the origin of the data, as long as these were not given by the person himself, and any existence of an automated decision-taking process, including profiling with meaningful information about the logic involved as well as the implications and intended effects of such procedures. Last but not least, if the personal data is transmitted to an unsecure third country, they must be informed of all suitable guarantees which were made.

Information can be transmitted to the impacted person as per Art. 12 para. 1 sentences 2 and 3 of the GDPR depending upon the facts in writing, electronically or verbally. According to the Art. 12(3) Information must be communicated quickly but at latest within 1 month. Only in justified exceptional cases may this 1-month deadline be exceeded. The information is, as a rule, given without payment. If, in addition, further copies are requested, one can request a reasonable payment which reflects administrative costs. In addition, the responsible party can also refuse granting information to an affected person in the case of unjustified or excessive requests. Responsible parties additionally have the right, if there is a large volume of information about the impacted person being processed, that they share their right of access regarding processing or information.

Suitable GDPR Articles
Art. 12 GDPR Transparent information, communication and modalities for the exercise of the rights of the data subject Art. 15 GDPR Right of access by the data subject Art. 46 GDPR Transfers subject to appropriate safeguards.

Right to Be Forgotten

The right to be forgotten derives from the case Google Spain SL, Google Inc. v Agencia Española de Protección de Datos, Mario Costeja González (2014). Now the right to be forgotten is being codified General Data Protection Regulation in addition to the right of erasure.

The correspondingly-named standard primarily regulates erasure obligations. According to this, personal data must be erased immediately as long as the data are no longer needed for their original processing purpose, or the impacted person has withdrawn his consent and there is no other reason for justification, the impacted person has objected and there is no preferential justified reason for the processing, or erasure is required to fulfill a statutory obligation under the EU law or the right of the Member States. In addition, data must of course be erased if the processing itself was against the law.

The responsible person is therefore subject on the one hand to automatic statutory erasure obligations, and must, on the other hand, comply with the impacted person's desire to be erased. The law does not further describe how the data must be erased in individual cases. The decisive element is that the result is that it is no longer possible to see the data without disproportionate expense. One regards this effort as sufficient if the media has been physically destroyed, or the data is permanently over-written using special software.

In addition, the right to be forgotten is found in Art. 17 para. 2 of the GDPR. If the responsible party has published the personal data, and if one of the above reasons to erasure is present, he must take suitable measures with consideration of the circumstances to inform all those who are further responsible for the data processing that all links to this personal data or copies or replicates of the personal data must be erased.

An application of erasure is not subject to any particular form, and the responsible party need not link it to such a form. However, the identity of the impacted person must be proven in a suitable way, as otherwise additional information could be requested from the responsible party, or the erasure could be refused. If there is an application to erase or a statutory obligation to erase, this must be implemented quickly. This means that the responsible party only has a suitable time to check the conditions for erasure. In the case of an application for erasure, the impacted party must be informed within 1 month about the measures taken or the reasons for refusal. Once again, the right to be forgotten is reflected in the obligation to notify. In addition to erasure, according to Art. 19 of the GDPR the responsible entity must inform all receivers of the data. For this, he must use all means available and exhaust all appropriate measures.

Suitable GDPR Articles

Art. 17 GDPR Right to erasure ('right to be forgotten') Art. 19 GDPR Notification obligation regarding rectification or erasure of personal data or restriction of processing.

Third Countries

It is essential these days to transmit data to third countries as part of international trade and cooperation. Examining the permissibility of such a transfer is done in two stages.

First, the data transfer itself must be permissible. Any processing of personal data is subject to a prohibition if permission is reserved. In addition to consent, Art. 6 of the GDPR sets forth further justification reasons, such as fulfilling a contract or protecting vital interests. For special personal data which requires a higher level of protection, the permission text of Art. 9 of the GDPR applies.

If the planned data transfer meets the general conditions, one must check in a second step whether transfer to the third country is permitted. One must differentiate between secure and unsecure third countries. Secure third countries are those for which the European Commission has confirmed a suitable level of protection in a decision of appropriateness. They provide in their national laws for protection of personal data which are comparable to those of EU law. At the time that the General Data Protection Regulation became applicable, the third countries which are counted as secure countries are: Andorra, Argentina, Canada (only commercial organizations), Faroe Islands, Guernsey, Israel, Isle of Man, Jersey, New Zealand, Switzerland, Uruguay and USA (if the receiver belongs to the Privacy Shield). Data transfer to these countries is expressly permitted.

If there is no decision of appropriateness for a country, this does not necessarily exclude data transfer to this country. Rather, the processor must ensure in another way that the personal data will be sufficiently protected by the recipient. This can be assured using standard data protection clauses, for data transfers within a Group through so-called "binding corporate rules," through obligation to comply with rules of behaviour which have been declared by the European Commission as being generally applicable, or by certification of the processing procedure.

Furthermore, there are several exceptions which legitimize data transfer to a third country, even if protection of personal data cannot be sufficiently assured. Most frequently, the consent of those impacted is relevant here. Thus, one must note the requirements for their voluntary nature. Further exceptions, such as transmitting to fulfill contracts, important reasons of public interest and the assertion of legal rights are less relevant in practice.

Suitable GDPR Articles
Art. 40 GDPR Codes of conduct Art. 42 GDPR Certification Art. 44 GDPR General principle for transfers Art. 45 GDPR Transfers on the basis of an adequacy decision Art. 46 GDPR Transfers subject to appropriate safeguards Art. 47 GDPR Binding corporate rules Art. 48 GDPR Transfers or disclosures not authorised by Union law Art. 49 GDPR Derogations for specific situations Art. 63 GDPR Consistency mechanism.

Appendix 3: U. S. Consumer Privacy Bill of Rights

112th Congress
1st Session
H. R. 1528
Consumer Privacy Protection Act[1]
To protect and enhance consumer privacy, and for other purposes.
IN THE HOUSE OF REPRESENTATIVES
April 13, 2011[2]
Mr. Stearns (for himself, Mr. Matheson, Mr. Bilbray, and Mr. Manzullo) introduced the following bill; which was referred to the Committee on Energy and Commerce
A BILL
To protect and enhance consumer privacy, and for other purposes.
Be it enacted by the Senate and House of Representatives of the United States of America in Congress assembled,

SEC. 1. Short Title

This Act may be cited as the "Consumer Privacy Protection Act of 2011".

---

[1] https://www.congress.gov/bill/112th-congress/house-bill/1528/text

[2] Unlike the success achieved in Europe to create a new and robust set of privacy legislation efforts in the United States have been frustrated by lobbying by the information and social media industries that derive tremendous amounts of income form trading in information about information. Grant Wythoff, "Silicon Valley is ready to make your decisions for you", *Washington Post,* August 19, 2018, Page B-3. Note that there is no "Section 2" in this draft legislation that never was passed into law in the United States.

© Springer Nature Switzerland AG 2019
J. N. Pelton, *Preparing for the Next Cyber Revolution,*
https://doi.org/10.1007/978-3-030-02137-5

SEC. 3. Definitions

In this Act, the following definitions apply:

(1) AFFILIATE.—The term "affiliate" means any company that controls, is controlled by, or is under common control with another company.

(2) COMMISSION.—The term "Commission" means the Federal Trade Commission.

(3) CONSUMER.—The term "consumer" means an individual acting in the individual's personal, family, or household capacity.

(4) COVERED ENTITY.— (A) The term "covered entity" means an entity (or an agent or affiliate of the entity) that collects (by any means, through any medium), sells, discloses for consideration, or uses personally identifiable information of more than 5000 consumers during any consecutive 12-month period, and includes a non-profit organization, including any organization described in section 501(c) of the Internal Revenue Code of 1986 that is exempt from taxation under section 501(a) of such Code, notwithstanding the definition of the term "Acts to regulate commerce" in section 4 of the Federal Trade Commission Act (15 U.S.C. 44) and the exception provided by section 5(a)(2) of such Act (15 U.S.C. 45(a)(2)) for such organizations.

(B) Such term does not include—

  (i) a governmental agency;
  (ii) a provider of professional services, or any affiliate thereof, to the extent that such provider is obligated by rules of professional ethics, or by applicable law or regulation, not to voluntarily disclose confidential client information without the consent of the client; or

(iii) a data processing outsourcing entity.

(5) DATA PROCESSING OUTSOURCING ENTITY.—The term "data processing outsourcing entity" means, with respect to a covered entity, a non-affiliated entity that—

(A) provides information technology processing, Web hosting, or telecommunications services to the covered entity;

(B) is contractually obligated to comply with security controls specified by the covered entity; and

(C) has no right to use the covered entity's personally identifiable information other than for performing data processing outsourcing services for the covered entity or as required by contract or law.

(6) DISPLAY.—The term "display" means intentionally communicating or otherwise making available (on the Internet or in any other manner) to another person.

(7) INFORMATION-SHARING AFFILIATE.—The term "information-sharing affiliate" means any affiliate that is under common control with a covered entity,

or is contractually obligated to comply with the practices enumerated under the privacy policy statement of the covered entity required under section 5.

(8) PERSONALLY IDENTIFIABLE INFORMATION.— (A) The term "personally identifiable information", with respect to a covered entity means individually identifiable information relating to a living individual who can be identified from that information, and includes:

(i) the combination of a first name (or initial) and last name of an individual, whether given at birth or time of adoption, or resulting from a lawful change of name;

(ii) the postal address of a physical place of residence of such individual;

(iii) an e-mail address of such individual;

(iv) a telephone number or mobile device number dedicated to contacting such individual at any place other than the individual's place of work;

(v) a social security number or other Federal or State government issued identification number issued to such individual; or

(vi) the complete account number of a credit or debit card issued to such individual.

(B) Such term also includes, when disclosed in connection with one or more of the items of information described in subparagraph (A)—

(i) a birth date, the number of a certificate of birth or adoption, or a place of birth; or

(ii) an electronic address, including an IP address.

(C) Such term does not include—

(i) anonymous or aggregate data, or any other information that does not identify a unique living individual;

(ii) information about a consumer inferred from data maintained about a consumer; or

(iii) information about a consumer that is publicly available or obtained from a public record.

(9) PROCESS.—The term "process", with respect to personally identifiable information, means any value-added activity performed on data by automated means.

(10) PUBLICLY AVAILABLE.—The term "publicly available", with respect to information, means information that is lawfully made available to the general public.

(11) PUBLIC RECORD.—The term "public record" means any item, collection, or grouping of information about an individual that is maintained by a Federal, State, or local government entity and that is made available to the public.

(12) PURCHASE.—The term "purchase" means providing, directly or indirectly, anything of value in exchange for a good or service.

(13) STATE.—The term "State" includes the several States, the District of Columbia, the Commonwealth of Puerto Rico, the Commonwealth of the

Northern Mariana Islands, American Samoa, Guam, the Virgin Islands, the Freely Associated States, and any other territory or possession of the United States.

(14) TRANSACTION.—The term "transaction" means an interaction between a consumer and a covered entity resulting in—

(A) any use of information that is necessary to complete the interaction in the course of which information is collected, or to maintain the provisioning of a good or service requested by the consumer, including use—

(i) to approve, guarantee, process, administer, complete, enforce, provide, or market a product, service, account, benefit, transaction, or payment method that is requested or approved by the consumer;

(ii) to deliver goods, services, funds, or other consideration to, or on behalf of, the consumer;

(iii) to protect the health and safety of the consumer; and

(iv) related to website analytics methods or measurements for improving or enhancing products or services.

(B) any disclosure of information that is necessary for the consumer to enforce any right of the consumer;

(C) any disclosure of information that is required by law or by a court order;

(D) any use of information to verify personally identifiable information by the consumer, evaluate, detect, or reduce the risk of fraud or other criminal activity, or other risk-management activities; and

(E) the collection or use of personally identifiable information for the marketing or advertising of a covered entity's products or services to its own customers or potential customers.

SEC. 4. Privacy Notices to Consumers

(a) Notice required.—A covered entity shall provide to a consumer a notice containing the information required under subsection (b) as follows:

(1) The covered entity shall provide the notice before any personally identifiable information that is collected from a consumer is used by the covered entity for a purpose unrelated to a transaction.

(2) Upon a material change in the covered entity's privacy policy under section 5(a), the covered entity shall provide the notice, not later than the first time after such change in policy that the covered entity seeks to sell, disclose for consideration, or use personally identifiable information to the extent practicable, to each consumer from whom the covered entity has collected such information.

(b) Form and contents of notice.—A notice required under subsection (a) shall be provided in a clear and conspicuous manner, be prominently displayed or explicitly stated to the consumer, and contain the following information:

(1) A statement that the personally identifiable information collected by the covered entity may be used or disclosed for purposes or transactions unrelated to that for which it was collected, as described in the covered entity's privacy statement.

(2) A description, appropriate to the applicable medium, of the manner in which the consumer may obtain a privacy policy statement that meets the requirements of section 5, which may include providing the consumer with an Internet website, a hyperlink to such a website, or a toll-free telephone number from which such a statement may be obtained. If the notice required under subsection (a) is provided to the consumer by means of an Internet website, one manner in which the consumer may obtain the privacy policy statement must be by means of an Internet website.

(3) If the notice is required under subsection (a)(2), a statement that there has been a material change in the covered entity's privacy policy.

SEC. 5. Privacy Policy Statements

(a) Privacy policy.—A covered entity shall establish a privacy policy with respect to the collection, sale, disclosure for consideration, dissemination, use, and security of the personally identifiable information of consumers, the principal elements of which shall be embodied in a privacy policy statement (or statements) that meets the requirements of subsection (b).

(b) Statement.—The statement (or statements) required under subsection (a) shall meet the following requirements:

(1) The statement must be brief, concise, clear, and conspicuous and written in plain language.

(2) The statement must be available to all consumers of the covered entity (regardless of the means by which a consumer conducts a transaction with the covered entity)—

(A) at no charge to the consumer; and

(B) at the time the covered entity first collects personally identifiable information about the consumer that may be used for a purpose unrelated to a transaction with the consumer and subsequently.

(3) The statement must disclose only the following:

(A) The identity of each covered entity, or a description of each class or type of covered entity, that may collect or use the information.

(B) The types of information that may be collected or used.

(C) How the information may be used.
(D) Whether the consumer is required to provide the information in order to do business with the covered entity.
(E) The extent to which the information is subject to sale or disclosure for consideration to a covered entity that is not an information-sharing affiliate of the covered entity providing the statement, including—

  (i) a clear and prominent statement of the fact that the information is subject to such sale or disclosure for consideration;
  (ii) a description of each class or type of covered entity to which the information may be sold or disclosed for consideration;
  (iii) to the extent practicable, the purpose for which the information may be used; and
  (iv) the types of information that may be sold or disclosed for consideration.

(F) Whether the information security practices of the covered entity meet the security requirements of section 8 in order to prevent unauthorized disclosure or release of personally identifiable information.

(c) Commission facilitation.—The Commission may take actions (including conducting industry-wide workshops) to facilitate the development of harmonized, universal wording or logo-based graphics in order to convey the contents of privacy policy statements required under this section.

SEC. 6. Consumer Opportunity to Limit Sale or Disclosure of Information

(a) Preclusion of sale or disclosure.—

  (1) REQUIREMENT.—A covered entity shall provide to the consumer, without charge, the opportunity to preclude any sale or disclosure for consideration of the consumer's personally identifiable information, provided in a particular data collection, that may be used for a purpose other than a transaction with the consumer, to any covered entity that is not an information-sharing affiliate of the covered entity providing such opportunity.
  (2) DURATION.—A preclusion on sale or disclosure for consideration of information established by a consumer under this subsection shall remain in effect for 5 years or until the consumer indicates otherwise, whichever occurs sooner. A covered entity may not seek reconsideration of a consumer's preclusion of such sale or disclosure until at least 1 year after such preclusion has been imposed by the consumer.

(b) Permission for sale or disclosure.—A covered entity may provide the consumer an opportunity to permit the sale or disclosure described in subsection (a)(1) in exchange for a benefit to the consumer.

(c) Accessibility.—The opportunity to preclude (or if offered, to permit) the sale or disclosure for consideration of information under this section must be both easy to access and use, and the notice of the opportunity to preclude must be clear and conspicuous.

SEC. 7. Consumer Opportunity to Limit Other Information Practices

If a covered entity provides to a consumer the opportunity to limit other practices of the covered entity with respect to a particular collection or use of personally identifiable information regarding the consumer, other than that required by section 6—

(1) a notice and description of such opportunity must appear in the privacy statement;
(2) such opportunity must be easy to access and to use; and
(3) any limitation exercised by the consumer pursuant to such opportunity shall remain in effect, unless—

    (A) the limitation is withdrawn by the consumer; or
    (B) the covered entity provides the consumer at least 30 days notice before materially changing the limitation or terminating its compliance with the limitation.

SEC. 8. Information Security Obligations

(a) Implementation.—A covered entity shall prepare, revise as necessary, and implement an information security policy that is applicable to the information security practices and treatment of personally identifiable information maintained by the covered entity, that is designed to prevent the unauthorized disclosure or release of such information.
(b) Management approval.—An information security policy created pursuant to paragraph (1) shall be considered and approved by the senior management officials of the covered entity.
(c) Contents.—An information security policy required under paragraph (1) shall include—

    (1) a process for taking corrective action to prevent or mitigate unauthorized disclosure of information; and
    (2) identifying an officer of the covered entity as the point of contact with responsibility for information security issues for the covered entity.

SEC. 9. Self-Regulatory Programs

(a) Self-Regulatory program.—

  (1) PRESUMPTION OF COMPLIANCE.—The Commission shall presume that a covered entity is in compliance with the provisions of sections 4 through 8 if that covered entity—

  (A) participates in a self-regulatory program approved under subsection (b); and

  (B) is subject to enforcement under a self-regulatory program's guidelines, procedures, requirements, and restrictions (including a remedial process under subsection (c)(7)).

  (2) EFFECT OF WILLFUL NONCOMPLIANCE.—A covered entity that participates in a self-regulatory program under this section shall not be liable for a civil penalty arising out of a violation of any provision of sections 4 through 8 unless such violation results from willful noncompliance with the guidelines, procedures, requirements, or restrictions of the program.

(b) Approval by Commission.—

  (1) APPROVAL.—The Commission shall, within 90 days after submission of an application for approval of a self-regulatory program under this section (or of a material change in a program previously approved by the Commission), approve such program (or change) if the Commission finds that the program (or change) complies with the requirements of subsection (c).

  (2) FORM OF APPLICATION.—The Commission shall accept an application for approval under paragraph (1) in any reasonable form the applicant may submit.

  (3) DURATION UNTIL RENEWAL.—A self-regulatory program approved by the Commission under paragraph (1) shall be approved for a period of 5 years.

  (4) REVOCATION OF APPROVAL.—The Commission may, after notice and opportunity for a hearing, revoke approval granted under paragraph (1), if the Commission finds that a self-regulatory program fails to meet the requirements of subsection (c).

  (5) JUDICIAL REVIEW.—Any order by the Commission denying approval of a self-regulatory program shall be subject to judicial review, as provided in section 706 of title 5, United States Code.

(c) Requirements of self-Regulatory program.—A self-regulatory program complies with the requirements of this subsection if the program provides each of the following:

(1) Guidelines and procedures requiring a program participant to provide substantially equivalent or greater protections for consumers and their personally identifiable information as are provided under sections 4 through 8.

(2) Procedures and requirements to provide for—

(A) an initial review of a participant's privacy statement and privacy policy, and subsequent review whenever such statement or policy is substantively changed;

(B) a participant's self-review and self-certification of its privacy policy and practices to ensure compliance with the guidelines, procedures, requirements, and restrictions of the program established under this subsection;

(C) a participant's subsequent periodic self-reviews and self-certifications, which shall occur at least annually, of the its privacy policy and practices to ensure continued compliance with such guidelines, procedures, requirements, and restrictions;

(D) submission of self-reviews and self-certifications under this paragraph to any administrator of the program; and

(E) random review of participants, which may concentrate on selected compliance issues, if the self-regulatory program conducts—

  (i) random compliance tests with respect to each participant not less frequently than every 3 years;

 (ii) a full compliance test of a particular participant in any case where noncompliance with any of the selected compliance issues has been identified; and

(iii) full compliance tests of participants with a high number of complaints against them.

(3) Procedures and requirements that ensure that a program participant provides a process for resolving disputes with consumers relating to the privacy policy and practices of the participant. Such dispute resolution process—

(A) must be available without charge to a consumer;

(B) must be available at a cost to the participant that is reasonable and does not discourage participation by the participant in such process;

(C) must ensure that consumers are informed of how to utilize the process;

(D) may include, as one choice among others, binding arbitration; and

(E) (i) must be completed within 60 days after submission of the dispute by the consumer; or

 (ii) must be completed within 90 days after submission of the dispute by the consumer, if the participant—

(I) determines that additional time is required to obtain information to make an informed decision with respect to the dispute; and

(II) notifies the consumer and the self-regulatory program that such additional time is required.

(4) Provisions for the use by participants in the program of a means (including the use of a seal) to represent the participant's participation in the program.

(5) With respect to any nonvoluntary suspension or termination of participation in the program because of the participant's failure to comply with the program, procedures or requirements to provide for the following:

(A) Publication of notice and the reasons for any such suspension or termination, except that no personally identifiable information related to such suspension or termination may be published.

(B) Notice to the Commission of any such termination.

(6) Requirements and restrictions that assure independence with respect to program eligibility, compliance, and dispute resolution mechanisms and decisions from improper interference by management or ownership of the self-regulatory program participant.

(7) A process for a noncompliant participant to take timely remedial action in order to come back into compliance with the program before suspension or termination of participation in the program.

(d) Consumer dispute resolution.—

(1) SELF-REGULATORY DISPUTE PROCESS.—If a consumer has a dispute with a participant in a self-regulatory program under this section or under section 5 of the Federal Trade Commission Act (15 U.S.C. 45) to the extent that such dispute pertains to the entity's privacy policy or practices required for participation in the self-regulatory program, the consumer shall initially seek resolution through the participant's dispute resolution process (established in accordance with subsection (c)(3)). The Commission shall promptly refer to the participant involved any dispute submitted to the Commission for which resolution has not been initially sought through such process.

(2) RESOLUTION BY COMMISSION.—A consumer may submit to the Commission for resolution a dispute with a participant in a self-regulatory program under this section, if the following requirements are met:

(A) The dispute was initially submitted under paragraph (1) for resolution through the participant's dispute resolution process.

(B) The dispute submitted under paragraph (1) is not resolved—

(i) within 60 days after submission of the dispute by the consumer; or

(ii) to the satisfaction of the consumer.

(C) Notice of the facts of the dispute is submitted to the Commission not later than 30 days after the date on which the consumer is notified of the resolution through the participant's dispute resolution process. (D) The consumer has not voluntarily accepted a resolution of the dispute under paragraph (1).

(D) The dispute was not resolved through binding arbitration.

(3) LIMITATION.—Nothing in this Act shall prevent the Commission from investigating compliance with this Act by a participant in a self-regulatory covered entity based upon a complaint from an individual or covered entity other than a consumer with a dispute with such participant, or on its own initiative, except that prior to instituting any such investigation the Commission shall afford the self-regulatory covered entity a reasonable opportunity to invoke its own remedial procedures and assure compliance by the participant.

(4) CLEAR AND CONVINCING EVIDENCE.—The presumption established by paragraph (1) of subsection (a) may be overcome by clear and convincing evidence of non-compliance.

(e) Nonrelease of certain information.—The Commission may not compel a participant in a self-regulatory program approved under subsection (b) (or an administrator of such a program) to provide proprietary information or personally identifiable information of consumers to the Commission unless the Commission provides assurances that such information will not be released to the public.

(f) Misrepresentation of self-Regulatory program participation.—It is unlawful for a covered entity to misrepresent that it is a participant in a self-regulatory program (including through any mechanism provided under subsection (c)(4)) when such covered entity is not, in fact, such a participant.

(g) Exempted entity participation.—An entity that is not a covered entity and that voluntarily participates in a self-regulatory program under this section shall enjoy the rights and benefits provided under this section in any action or investigation under section 5 of the Federal Trade Commission Act (15 U.S.C. 45) to the extent that such action or investigation pertains to the entity's privacy policy or practices required for participation in the self-regulatory program.

SEC. 10. Enforcement

(a) Unfair or deceptive Act or practice.—A violation of any provision of this Act by a covered entity is an unfair or deceptive act or practice unlawful under section 5(a)(1) of the Federal Trade Commission Act (15 U.S.C. 45(a)(1)), except that the amount of any civil penalty under such Act shall be doubled for a violation of this Act, but may not exceed $500,000 for all related violations by a single violator (without respect to the number of consumers affected or the duration of the related violations).

(b) Guidelines and opinions.—In order to assist in compliance with this Act, the Federal Trade Commission may promulgate regulations and interpretive rules under section 18 of the Federal Trade Commission Act (15 U.S.C. 57a), with respect to specific types of acts or practices that would, or would not, comply with this Act.

SEC. 11. No Private Right of Action

This Act may not be considered or construed to provide any private right of action. No private civil action relating to any act or practice governed under this Act may be commenced or maintained in any State court or under State law (including a pendent State claim to an action under Federal law).

SEC. 12. Effect on Other Laws

(a) Qualified exemption for compliance with other Federal privacy laws.—To the extent that personally identifiable information protected under this Act is also protected under a provision of Federal privacy law described in subsection (c), a covered entity that complies with the relevant provision of such other Federal privacy law shall be deemed to have complied with the corresponding provision of this Act.

(b) Protection of other Federal privacy laws.—Nothing in this Act may be construed to modify, limit, supersede, or interfere with the operation of the Federal privacy laws described in subsection (c) or the provision of information permitted or required, expressly or by implication, by such laws, with respect to Federal rights and practices.

(c) Other Federal privacy laws described.—The provisions of law to which subsections (a) and (b) apply are the following:

    (1) Section 552a of title 5, United States Code (commonly known as the Privacy Act of 1974).

    (2) The Right to Financial Privacy Act of 1978 (12 U.S.C. 3401 et seq.).

    (3) The Fair Credit Reporting Act (15 U.S.C. 1681 et seq.).

    (4) The Fair Debt Collection Practices Act (15 U.S.C. 1692 et seq.).

    (5) The Children's Online Privacy Protection Act of 1998 (15 U.S.C. 6501 et seq.).

    (6) Title V of the Gramm-Leach-Bliley Act of 1999 (15 U.S.C. 6801 et seq.).

    (7) The Electronic Communications Privacy Act of 1986 (Public Law 99–508).

    (8) The Driver's Privacy Protection Act of 1994 (18 U.S.C. 2721 et seq.).

    (9) The Family Educational Rights and Privacy Act of 1974 (20 U.S.C. 1221 note, 1232 g).

    (10) Section 445 of the General Education Provisions Act (20 U.S.C. 1232 h).

    (11) The Privacy Protection Act of 1980 (42 U.S.C. 2000aa et seq.).

    (12) Section 222 of the Communications Act of 1934 (47 U.S.C. 222) relating to the Customer Proprietary Network Information.

    (13) The Cable Communications Policy Act of 1984 (47 U.S.C. 521 et seq.).

    (14) The Communications Assistance for Law Enforcement Act (47 U.S.C. 1001 et seq.).

(15) The Video Privacy Protection Act of 1988 (Public Law 100–618).

(16) The Telephone Consumer Protection Act of 1991 (Public Law 102–243).

(17) The Health Insurance Portability and Accountability Act of 1996 (Public Law 104–191), as it relates to an entity described in section 1172(a) of the Social Security Act (42 U.S.C. 1320d–1(a)) or to activities regulated under section 1173 of such Act (42 U.S.C. 1320d–2).

(18) The CAN–SPAM Act of 2003 (15 U.S.C. 7701 et seq.).

(d) Preemption of State privacy laws.—This Act preempts any statutory law, common law, rule, or regulation of a State, or a political subdivision of a State, to the extent such law, rule, or regulation relates to or affects the collection, use, sale, disclosure, retention, or dissemination of personally identifiable information in commerce. No State, or political subdivision of a State, may take any action to enforce this Act.

SEC. 13. Effective Date

This Act shall apply with respect to personally identifiable information collected on or after the date that is 1 year after the date of enactment of this Act.

Glossary

3-D manufacturing This is the part of the additive manufacturing where objects are created using 3-D printers and unique printing materials to create new industrial products in a software-defined way rather than traditional manufacturing using mills, lathes or carving devices.

Additive manufacturing This is a revolutionary form of computer driven manufacturing that is also sometimes referred to as 3-D manufacturing. This type of manufacturing depends on the use of data provided via computer-aided design (CAD) software or via 3-D object scanners. This software is then used to direct hardware to deposit material in an 'additive' fashion on a layer on layer basis in order to create precise geometric shapes. Thus the additive manufacturing process adds material to create an object using a software-driven design. This is the opposite of traditional manufacturing, which 'subtracts' materials to create a manufactured object by machining, carving, or other similar activities.

AI Artificial Intelligence (see below).

Anthropocene Age This is the current geological age. Scientists have collectively decided that this is the most apt descriptive designation to apply to Earth in its current stage of evolution. Ir is based on their collective view that humanity is the primary force that is shaping the world today.

Anthropocene extinction This phrase is now applied by some that fear that human consumption of resources, human-fueled climate change and other similar concerns could lead to mass extinction of many species (flora and fauna) that will accompany runaway climate change. This is also sometimes described as 'the sixth extinction' or the 'sixth mass extinction.' The so-called K-T mass extinction was the fifth, which led to the extinction of up to about 80% of all species on Earth.

Artificial intelligence (AI) This phrase suggests a higher level of logical machine analysis. The process transcends a set of predetermined algorithms with if-then choices that lead to fixed outcomes. This is type of higher machine logic is sometimes referred to a heuristic algorithm.

© Springer Nature Switzerland AG 2019
J. N. Pelton, *Preparing for the Next Cyber Revolution*,
https://doi.org/10.1007/978-3-030-02137-5

Bot This can be a smart device or highly capable software that can execute commands, reply to messages, or perform routine tasks, such as online searches, either automatically or with little or no human intervention. The word bot derives from a shortened form of robot, with the distinction here being that a bot can either be software or a device with controlling software.

Circular economy This is a concept of developing a 'sustainable' economic system that is based on recycling of resources and avoiding a process with higher and higher rates of throughput based on disposing of products and resources rather than recycling them. (See also **Disposable economy**.)

Climate change This is the systematic set of changes to Earth's climate over time that is leading to a wide range of important variations in the world's atmosphere, ocean, storm and rain patterns, areas of drought and aridity, etc. These changes to the climate appear to be primarily driven by human activities, economic systems, and lack of sustainable practices.

Consumer Privacy Bill of Rights This is a U. S. initiative that seeks to ensure greater protection of U. S. citizen privacy and is in some senses of the word a counterpart to the European General Data Protection Regulations that entered into force in 2018.

Cyber-criminals These are hackers, crackers, and other types of cyber-based criminals that use the Internet and other digital means so as to extort information, obtain stolen personal information via the dark web and many other illegal means using surreptitious software or other digital means in order to carry out criminal attacks.

Cybernetics The science or discipline of understanding, defining, and improving systems by obtaining feedback from them. This particularly applies to digitally networked systems and the software that controls them. A circular process, it involves causal chains and uses feedback to achieve more efficient and accurate control of these systems with ever greater effectiveness.

Cybersecurity This is the effort to prevent technology-related crimes using various techniques, tools and software especially as applied to the protection of digital networks and the privacy of users of digital networks.

Disposable economy This is the type of economic growth that is most generally prevalent in today's developed world and is symptomatic of conventional capitalist growth practices. The concept is based on ever expanding, exponential growth of human markets that is based on consumption of resources, devices and products that then replaced by an ever increasing cycle of consumption of new products manufactured out of newly mined or processes resources. It places no particular value on recycling, sustainability of markets or economic systems for the longer term.

Disruptive technology This is a revolutionary technology that envisions an entirely new way of producing a product (i.e., additive manufacturing) or a new process or technique to deliver a service (i.e., Uber or Airbnb).

European General Data Protection Regulations This is a new set of digital protection regulations for citizens that went into effect in May 2018 under the auspices of the European Union. Since most global businesses with digital operations conduct business in Europe these new regulations have had a global impact. (Also called the **General Data Protection Regulations.**)

Future compression This a term developed by the author to describe the increasing rate of technological progress and exponential creation of new information and knowledge in modern society.

Hyper-object This is a term coined by the environmentalist Tim Morton. It refers to large-scope events, transitions or concepts that generally develop over a long period of time. Concepts such as climate change, the Anthropocene Age, super-automation and indeed the entire digital age are hyper-objects. These powerful and sustained events are so profound in nature, long term in effect, and transformational in an economic, social, and technical sense so that the typical citizen feels that they are beyond their own comprehension and certainly beyond the scope of their own abilities to steer or revise their impact.

Internet of Everything (IoE) This is foreseen as the next step beyond the Internet of Things (see next). This concept envisions more and more autonomous functionality of all types of manufactured objects that become more self-aware and are able to communicate via digital networks about all aspects of their functionality.

Internet of Things (IoT) This is a network of physical devices, such as vehicles, home appliances, and other items embedded with electronics, software, sensors, and more that can connect and exchange data using machine-to-machine communications. It began with the use of RFID (radio frequency ID) devices used for shipping and inventory purposes, but has grown to include equipping automobiles, trucks, appliances and all sorts of devices.

Meta-cities These are smaller cities surrounding a megacity (city with a huge population) that is in danger of becoming too big to be functional.

NewSpace This refers to the current entrepreneurial developments in the space industry to include new disruptive technologies and new ways to create more cost effective and more competitive technologies and systems in areas such as the design of launch vehicles, small satellites and large-scale satellite constellations, and new ways of doing business. Many of these innovations come from the computer and IT industries meeting the conventional aerospace industry. This series of innovations that are sometimes described as Silicon Valley meets the aerospace industry is also referred to as Space 2.0.

Pattern Recognition Theory of Mind (PRTM) This is a concept primarily developed by Ray Kurzweil as to how the human cortex operates and that human thinking is primarily based on hierarchical pattern recognition. The cortex can store many more patterns than today's fastest computer, but computers can be compensate by becoming much faster in terms of petaflops of processing power. One of the steps forward to achieving the Singularity (see below) and super-automation appears to be based on developing thinking machines that can use a similar hierarchical pattern recognition process that operates similar to the human neo-cortex, as argued by Ray Kurzweil.

Smart city This refers the core concepts related to the development of essential functionalities and design aspects of ideal cities for the future, using AI technologies and big data, among other things. Thus the elements of a smart city will continue to evolve over time to become more responsive to the needs of city residents as we progress into the future.

Smart robotics This refers to the develop of new machines with increasingly powerful processing capabilities based on hierarchical pattern recognition so that machines have a higher and higher level of artificial intelligence (AI) and are able to perform more and more tasks in a modern economy. Such smart robotics can perform their tasks 24 h a day and 7 days a week without stopping for food, bathroom breaks or seeking higher wages and better benefit compensation. Ray Kurzweil has predicted that truly smart robots can be achieved by around 2030.

Space 2.0 This is the same concept as NewSpace, which was defined earlier.

Super-automation This term refers to the development of smarter and smarter robots and software with artificial intelligence and heuristic algorithms that can perform more and more tasks within modern society and carry out all types of tasks, including sophisticated professional jobs now requiring advanced education and that this will impact all aspects of society in a profound way.

Techno-terrorist These are hackers or cyber-criminals who carry out terrorist attacks against countries, vital infrastructure, and key institutions. The latest iteration of techno-terrorist attacks not only involve attacks on physical infrastructure such as power plants, nuclear reactors, transportation systems, etc., but also attacks against financial institutions and political elections.

The Fourth Wave The first wave in human history began with the shift from being hunter/gatherers to farmers. Then there was the shift from farming to industrial production or the Industrial Age – the Second Wave. The Third Wave was the shift to a post-industrial economy. This largely happened in the 1950s, as prime employment shifted from industry and manufacturing to a service economy. The Fourth Wave will be fueled by super-automation, wherein humans will be increasingly partnered with intelligent machines, smart robots and the post-services economy. This will create a major upheaval in jobs and employment and see shifts in the economy to new types of economic mechanisms such as universal basic income payments (see later).

The Singularity This is the future situation predicted by Ray Kurzweil in his book *The Singularity is Near* that might come as early as 2030. In this new economic and social environment smart robots and AI software systems using heuristic algorithms will have the equivalent reasoning power and smarts to carry out a large portion of today's industrial and services jobs.

Troll This is someone that uses the Internet and particularly social media sites to attack, make fun of, deride, or otherwise spread malicious information or bully other people.

U. N. Sustainable Development Goals for 2030/U. N. Millennium Goals The U. N. General Assembly agreed to a series of goals for global development at the time of the millennium starting in 2001. These goals were set to be accomplished by 2015. In 2015 the United Nations\developed 17 sustainable development goals that are now hoped to be accomplished by 2030. These current goals are provided as an appendix to this book. New cyber-technologies, Space

2.0 capabilities and other advanced systems represent some of the tools to help achieve those goals. Ironically these same tools also pose issues and problems to be overcome with regard to other goals cited by the General Assembly.

Universal Basic Income (UBI) payments Some have argued that one of the consequences of super-automation and the Singularity is a lack of availability of the traditional forms of employment and that the current economic systems of many countries will have to adjust by creating for all people within a society a Universal Basic Income (UBI) payment, or salary to sustain them. This is because smart robots will do most of the farming, mining, and manufacturing as well as service jobs.

Index

© Springer Nature Switzerland AG 2019
J. N. Pelton, *Preparing for the Next Cyber Revolution*,
https://doi.org/10.1007/978-3-030-02137-5